MISSILE DEFENSE

Other books in the At Issue series:

MISSILE DEFENSE

William Dudley, *Book Editor*

Daniel Leone, *President*
Bonnie Szumski, *Publisher*
Scott Barbour, *Managing Editor*

GREENHAVEN
PRESS ®

San Diego • Detroit • New York • San Francisco • Cleveland
New Haven, Conn. • Waterville, Maine • London • Munich

LIBRARY OF CONGRESS CATALOGING-IN-PUBLICATION DATA

Missile defense / William Dudley, book editor.
    p. cm. — (At issue)
  Includes bibliographical references and index.
  ISBN 0-7377-1329-1 (pbk. : alk. paper) — ISBN 0-7377-1328-3 (lib. : alk. paper)
    1. Ballistic missile defenses—United States. I. Dudley, William, 1964– .
  II. At issue (San Diego, Calif.)
  UG743 .M577 2003
  358.1'74'0973—dc21   2002029804

Printed in the United States of America

Contents

Introduction

On June 15, 2002, groundbreaking ceremonies were held at a test site in Delta Junction, Alaska, to celebrate a $64 billion program to develop a national missile defense (NMD) system for the United States. The ceremony also marked the end of a diplomatic era by signifying America's withdrawal from the Anti-Ballistic Missile (ABM) Treaty, a thirty-year-old pact between the United States and the Soviet Union (later Russia) that had banned such missile defense development. The termination of the ABM Treaty was hailed by some observers as a welcome step toward defending America from missile attack. For others, the end of the treaty signified a step toward a dangerous new arms race.

In 1972, when the ABM Treaty was negotiated, the United States and the Soviet Union were the world's dominant superpowers. They were locked in longstanding "Cold War" hostilities toward each other, which had resulted in a nuclear arms race. By 1972 both countries had thousands of missiles aimed at each other, each missile armed with one or more nuclear warheads much more destructive than the weapons that destroyed Hiroshima and Nagasaki in 1945. The opposing arsenals were powerful enough to destroy each country many times over. The reason both countries developed and deployed so many seemingly redundant bombs was rooted in the logic of nuclear deterrence, sometimes referred to as MAD (mutually assured destruction). The theory was that both sides would refrain from actually using these weapons in order to avoid being destroyed by the other country's nuclear response.

The ABM Treaty, which was signed by American president Richard Nixon and Soviet leader Leonid Brezhnev in May 1972, was also to some extent an outgrowth of MAD. The goal was to prevent a defensive arms race as well as maintain each side's nuclear deterrent over the other. If one side developed a system that prevented enemy missiles from reaching their targets, nuclear deterrence would no longer apply and a nuclear war might result. The ABM Treaty did not totally ban missile defense but it severely limited the number and range of defensive missiles each country could deploy.

For thirty years the ABM Treaty—viewed by many observers as the cornerstone of arms control between the two superpowers—was the diplomatic backdrop to debates within the United States over developing missile defense. It figured prominently in the debate over President Ronald Reagan's 1983 Strategic Defense Initiative (SDI). Reagan's plan, dubbed "Star Wars" by the media, proposed an array of satellite-based and other weapons that would intercept and destroy enemy missiles and make the Soviet nuclear arsenal, in Reagan's words, "impotent and obsolete." However, his proposal was strongly criticized by members of Congress and others as reckless and unworkable—and a violation of the ABM Treaty to boot. Reagan's successors, George H.W. Bush and Bill Clinton, continued to fund some missile defense research but kept such research within the limits of the ABM Treaty.

George W. Bush takes office

When George W. Bush became president in 2001, the cause of missile defense was significantly boosted. Bush argued that the ABM Treaty had outlived its usefulness and was an unnecessary roadblock to national missile defense. The world had changed since 1972, Bush and his supporters argued, noting the fact that the Cold War ended in 1991 when the Soviet Union collapsed. The republic of Russia, which inherited most of the Soviet Union's nuclear arsenal and its international obligations including the ABM Treaty, has agreed to a series of cutbacks on its nuclear forces, making it less of a threat to America.

The leading nuclear threats now facing the United States, Bush and others contend, are countries such as North Korea, Iraq, or Iran who harbor hostilities toward the United States and are suspected of developing nuclear weapons. Nuclear deterrence cannot be counted on to deter such "rogue" states, Bush and his defenders maintain. "Deterrence can no longer be based solely on the threat of nuclear retaliation," Bush declared in a May 2001 speech. "Today's world requires a new policy."

In December 2001 Bush formally gave Russia and the world a six-month notice (provided for in the pact) that the United States would withdraw from the ABM Treaty in order to pursue rapid deployment of a missile defense system. Unlike Reagan's SDI, the missile defense system proposed by the Bush administration would protect U.S. allies and deployed forces in addition to the United States itself.

Bush's decision to abrogate the ABM Treaty removed one major obstacle to the development of an American missile defense system. Other significant obstacles remain, however, including technological hurdles. Opponents of missile defense maintain that the task of intercepting and destroying incoming ballistic missiles is a formidable one—and that letting even one nuclear missile slip by is an unacceptable risk. Other potential obstacles to missile defense include the reaction of other countries. Critics argue that an aggressive U.S. pursuit of missile defense risks alienating many countries (including American allies), and might provoke China or Russia or other nations to develop more sophisticated offensive weapons.

The terrorist attacks of September 11, 2001, added new elements to the debate about national missile defense. Proponents argue that the attacks justify developing a missile defense system because they illustrate the gravity of the threat that terrorists pose to America—a threat that might in the future include terrorism involving chemical, biological, or nuclear weapons. Skeptics argue that a missile defense system as advocated by Bush would not have prevented the September 11 incidents—nor could it prevent a terrorist group from finding alternatives to missiles to deliver weapons of mass destruction.

America's determined pursuit of missile defense, as indicated by Bush's decision to shelve the ABM Treaty, guarantees that the issues surrounding it will continue to be debated. The articles in this volume present a spectrum of viewpoints on the feasibility and desirability of national missile defense.

1

The Need for Missile Defense Is More Pressing than Ever

George W. Bush

George W. Bush is president of the United States. The following is taken from a May 2001 speech to students and faculty at National Defense University.

The world has changed significantly since the United States and the Soviet Union signed the Anti-Ballistic Missile (ABM) Treaty in 1972. The Cold War is over, but the world remains a dangerous place in which many countries, some with irresponsible leaders, have nuclear weapons and ballistic missile capabilities. The United States can no longer rely on the deterrence threat of nuclear retaliation to protect itself and its interests. America needs to pursue technologies that can protect against missile attacks.

I want us to think back some 30 years to a far different time in a far different world. The United States and the Soviet Union were locked in a hostile rivalry. The Soviet Union was our unquestioned enemy; a highly-armed threat to freedom and democracy. Far more than that wall in Berlin divided us.

Our highest ideal was—and remains—individual liberty. Theirs was the construction of a vast communist empire. Their totalitarian regime held much of Europe captive behind an iron curtain.

We didn't trust them, and for good reason. Our deep differences were expressed in a dangerous military confrontation that resulted in thousands of nuclear weapons pointed at each other on hair-trigger alert. Security of both the United States and the Soviet Union was based on a grim premise: that neither side would fire nuclear weapons at each other, because doing so would mean the end of both nations.

We even went so far as to codify this relationship in a 1972 ABM [Anti-Ballistic Missile] Treaty, based on the doctrine that our very survival would best be insured by leaving both sides completely open and vulner-

Excerpted from "Remarks by the President to Students and Faculty at National Defense University," by George W. Bush, www.whitehouse.gov, May 1, 2001.

9

able to nuclear attack. The threat was real and vivid. The Strategic Air Command had an airborne command post called the Looking Glass, aloft 24 hours a day, ready in case the President ordered our strategic forces to move toward their targets and release their nuclear ordnance.

The Soviet Union had almost 1.5 million troops deep in the heart of Europe, in Poland and Czechoslovakia, Hungary and East Germany. We used our nuclear weapons not just to prevent the Soviet Union from using their nuclear weapons, but also to contain their conventional military forces, to prevent them from extending the Iron Curtain into parts of Europe and Asia that were still free.

In that world, few other nations had nuclear weapons and most of those who did were responsible allies, such as Britain and France. We worried about the proliferation of nuclear weapons to other countries, but it was mostly a distant threat, not yet a reality.

A changed world

Today, the sun comes up on a vastly different world. The Wall is gone, and so is the Soviet Union. Today's Russia is not yesterday's Soviet Union. Its government is no longer Communist. Its president is elected. Today's Russia is not our enemy, but a country in transition with an opportunity to emerge as a great nation, democratic, at peace with itself and its neighbors. The Iron Curtain no longer exists. Poland, Hungary and the Czech Republic are free nations, and they are now our allies in NATO, together with a reunited Germany.

Yet, this is still a dangerous world, a less certain, a less predictable one. More nations have nuclear weapons and still more have nuclear aspirations. Many have chemical and biological weapons. Some already have developed the ballistic missile technology that would allow them to deliver weapons of mass destruction at long distances and at incredible speeds. And a number of these countries are spreading these technologies around the world.

We need a new framework that allows us to build missile defenses to counter the different threats of today's world.

Most troubling of all, the list of these countries includes some of the world's least-responsible states. Unlike the Cold War, today's most urgent threat stems not from thousands of ballistic missiles in the Soviet hands, but from a small number of missiles in the hands of these states, states for whom terror and blackmail are a way of life. They seek weapons of mass destruction to intimidate their neighbors, and to keep the United States and other responsible nations from helping allies and friends in strategic parts of the world.

When Iraqi president Saddam Hussein invaded Kuwait in 1990, the world joined forces to turn him back. But the international community would have faced a very different situation had Hussein been able to blackmail with nuclear weapons. Like Saddam Hussein, some of today's tyrants are gripped by an implacable hatred of the United States of America. They hate our friends, they hate our values, they hate democracy and freedom and individual liberty. Many care little for the lives of their own people. In such a world, Cold War deterrence is no longer enough.

When ready, and working with Congress, we will deploy missile defenses to strengthen global security and stability.

To maintain peace, to protect our own citizens and our own allies and friends, we must seek security based on more than the grim premise that we can destroy those who seek to destroy us. This is an important opportunity for the world to re-think the unthinkable, and to find new ways to keep the peace.

Today's world requires a new policy, a broad strategy of active nonproliferation, counterproliferation and defenses. We must work together with other like-minded nations to deny weapons of terror from those seeking to acquire them. We must work with allies and friends who wish to join with us to defend against the harm they can inflict. And together we must deter anyone who would contemplate their use.

We need new concepts of deterrence that rely on both offensive and defensive forces. Deterrence can no longer be based solely on the threat of nuclear retaliation. Defenses can strengthen deterrence by reducing the incentive for proliferation.

A new framework

We need a new framework that allows us to build missile defenses to counter the different threats of today's world. To do so, we must move beyond the constraints of the 30 year old ABM Treaty. This treaty does not recognize the present, or point us to the future. It enshrines the past. No treaty that prevents us from addressing today's threats, that prohibits us from pursuing promising technology to defend ourselves, our friends and our allies is in our interests or in the interests of world peace.

This new framework must encourage still further cuts in nuclear weapons. Nuclear weapons still have a vital role to play in our security and that of our allies. We can, and will, change the size, the composition, the character of our nuclear forces in a way that reflects the reality that the Cold War is over.

I am committed to achieving a credible deterrent with the lowest-possible number of nuclear weapons consistent with our national security needs, including our obligations to our allies. My goal is to move quickly to reduce nuclear forces. The United States will lead by example to achieve our interests and the interests for peace in the world.

Several months ago, I asked Secretary of Defense Donald Rumsfeld to examine all available technologies and basing modes for effective missile defenses that could protect the United States, our deployed forces, our friends and our allies. The Secretary has explored a number of complementary and innovative approaches.

The Secretary has identified near-term options that could allow us to deploy an initial capability against limited threats. In some cases, we can draw on already established technologies that might involve land-based and sea-based capabilities to intercept missiles in mid-course or after they re-enter the atmosphere. We also recognize the substantial advantages of intercepting missiles early in their flight, especially in the boost phase.

The preliminary work has produced some promising options for advanced sensors and interceptors that may provide this capability. If based at sea or on aircraft, such approaches could provide limited, but effective, defenses.

We have more work to do to determine the final form the defenses might take. We will explore all these options further. We recognize the technological difficulties we face and we look forward to the challenge. Our nation will assign the best people to this critical task.

We will evaluate what works and what does not. We know that some approaches will not work. We also know that we will be able to build on our successes. When ready, and working with Congress, we will deploy missile defenses to strengthen global security and stability.

Consulting with allies

I've made it clear from the very beginning that I would consult closely on the important subject with our friends and allies who are also threatened by missiles and weapons of mass destruction. . . .

These will be real consultations. We are not presenting our friends and allies with unilateral decisions already made. We look forward to hearing their views, the views of our friends, and to take them into account.

We will seek their input on all the issues surrounding the new strategic environment. We'll also need to reach out to other interested states, including China and Russia. Russia and the United States should work together to develop a new foundation for world peace and security in the 21st century. We should leave behind the constraints of an ABM Treaty that perpetuates a relationship based on distrust and mutual vulnerability. This Treaty ignores the fundamental breakthroughs in technology during the last 30 years. It prohibits us from exploring all options for defending against the threats that face us, our allies and other countries.

That's why we should work together to replace this Treaty with a new framework that reflects a clear and clean break from the past, and especially from the adversarial legacy of the Cold War. This new cooperative relationship should look to the future, not to the past. It should be reassuring, rather than threatening. It should be premised on openness, mutual confidence and real opportunities for cooperation, including the area of missile defense. It should allow us to share information so that each nation can improve its early warning capability, and its capability to defend its people and territory. And perhaps one day, we can even cooperate in a joint defense.

I want to complete the work of changing our relationship from one based on a nuclear balance of terror, to one based on common responsibilities and common interests. We may have areas of difference with Russia, but we are not and must not be strategic adversaries. Russia and America both face new threats to security. Together, we can address today's threats and pursue today's opportunities. We can explore technologies that have the potential to make us all safer.

This is a time for vision; a time for a new way of thinking; a time for bold leadership. The Looking Glass no longer stands its 24-hour-a-day vigil. We must all look at the world in a new, realistic way, to preserve peace for generations to come.

2

America Should Pursue a Ballistic Missile Defense System

Brian T. Kennedy

Brian T. Kennedy is a vice president of the Claremont Institute, a public policy research foundation.

If a nation or terrorist group decided to attack the United States with missiles, America would be unable to defend itself and would suffer helplessly as millions were killed. Moreover, other nations, such as Russia or China, could use the threat of a missile attack to force the United States into abandoning its allies and strategic interests. The United States has the technological capabilities to develop a ballistic missile defense system. America's leaders should make missile defense a leading national priority.

On September 11 [2001], our nation's enemies attacked us using hijacked airliners. Next time, the vehicles of death and destruction might well be ballistic missiles armed with nuclear, chemical, or biological warheads. And let us be clear: The United States is defenseless against this mortal danger. We would today have to suffer helplessly a ballistic missile attack, just as we suffered helplessly on September 11. But the dead would number in the millions and a constitutional crisis would likely ensue, because the survivors would wonder—with good reason—if their government were capable of carrying out its primary constitutional duty: to "provide for the common defense."

The threat is real

The attack of September 11 should not be seen as a fanatical act of individuals like Osama Bin Laden, but as a deliberate act of a consortium of nations who hope to remove the U.S. from its strategic positions in the Middle East, in Asia and the Pacific, and in Europe. It is the belief of such nations that the U.S. can be made to abandon its allies, such as Israel, if the cost of standing by them becomes too high. It is not altogether un-reasonable for our enemies to act on such a belief. The failure of U.S. po-litical leadership, over a period of two decades, to respond proportion-ately to terrorist attacks on Americans in Lebanon, to the first World Trade Center bombing, to the attack on the Khobar Towers in Saudi Ara-bia, to the bombings of U.S. embassies abroad, and most recently to the attack on the USS *Cole* in Yemen, likely emboldened them. They may also have been encouraged by observing our government's unwillingness to defend Americans against ballistic missiles. For all of the intelligence fail-ures leading up to September 11, we know with absolute certainty that various nations are spending billions of dollars to build or acquire strate-gic ballistic missiles with which to attack and blackmail the United States. Yet even now, under a president who supports it, missile defense ad-vances at a glacial pace.

Who are these enemy nations, in whose interest it is to press the U.S. into retreating from the world stage? Despite the kind words of Russian President Vladimir Putin, encouraging a "tough response" to the terrorist attack of September 11, we know that it is the Russian and Chinese gov-ernments that are supplying our enemies in Iraq, Iran, Libya, and North Korea with the ballistic missile technology to terrorize our nation. Is it possible that Russia and China don't understand the consequences of transferring this technology? Are Vladimir Putin and Jiang Zemin un-aware that countries like Iran and Iraq are known sponsors of terrorism? In light of the absurdity of these questions, it is reasonable to assume that Russia and China transfer this technology as a matter of high government policy, using these rogue states as proxies to destabilize the West because they have an interest in expanding their power, and because they know that only the U.S. can stand in their way.

Nations are spending billions of dollars to build or acquire strategic ballistic missiles with which to attack and blackmail the United States.

We should also note that ballistic missiles can be used not only to kill and destroy, but to commit geopolitical blackmail. In February of 1996, during a confrontation between mainland China and our democratic ally on Taiwan, Lieutenant General Xiong Guang Kai, a senior Chinese offi-cial, made an implicit nuclear threat against the U.S., warning our gov-ernment not to interfere because Americans "care more about Los Ange-les than they do Taipei." With a minimum of 20 Chinese intercontinental ballistic missiles (ICBMs) currently aimed at the U.S., such threats must be taken seriously.

The strategic terror of ballistic missiles

China possesses the DF-5 ballistic missile with a single, four-megaton war-head. Such a warhead could destroy an area of 87.5 square miles, or roughly all of Manhattan, with its daily population of three million people. Even more devastating is the Russian SS-18, which has a range of

7,500 miles and is capable of carrying a single, 24-megaton warhead or multiple warheads ranging from 550 to 750 kilotons.

Imagine a ballistic missile attack on New York or Los Angeles, resulting in the death of three to eight million Americans. Beyond the staggering loss of human life, this would take a devastating political and economic toll. Americans' faith in their government—a government that allowed such an attack—would be shaken to its core. As for the economic shock, consider that damages from the September 11 attack, minor by comparison, are estimated by some economists to be nearly 1.3 trillion dollars, roughly one-fifth of GNP.

It would be better to prevent a nuclear attack than to suffer one and retaliate.

Missile defense critics insist that such an attack could never happen, based on the expectation that the U.S. would immediately strike back at whomever launched it with an equal fury. They point to the success of the Cold War theory of Mutually Assured Destruction (MAD). But even MAD is premised on the idea that the U.S. would "absorb" a nuclear strike, much like we "absorbed" the attack of September 11. Afterwards the President, or surviving political leadership, would estimate the losses and then employ our submarines, bombers, and remaining land-based ICBMs to launch a counterattack. This would fulfill the premise of MAD, but it would also almost certainly guarantee additional ballistic missile attacks from elsewhere.

Consider another scenario. What if a president, in order to avoid the complete annihilation of the nation, came to terms with our enemies? What rational leader wouldn't consider such an option, given the unprecedented horror of the alternative? Considering how Americans value human life, would a Bill Clinton or a George Bush order the unthinkable? Would *any* president launch a retaliatory nuclear strike against a country, even one as small as Iraq, if it meant further massive casualties to American citizens? Should we not agree that an American president ought not to have to make such a decision? President Ronald Reagan expressed this simply when he said that it would be better to prevent a nuclear attack than to suffer one and retaliate.

Then there is the blackmail scenario. What if Osama Bin Laden were to obtain a nuclear ballistic missile from Pakistan (which, after all, helped to install the Taliban regime), place it on a ship somewhere off our coast, and demand that the U.S. not intervene in the destruction of Israel? Would we trade Los Angeles or New York for Tel Aviv or Jerusalem? Looked at this way, nuclear blackmail would be as devastating politically as nuclear war would be physically.

Roadblock to defense: the ABM Treaty

Signed by the Soviet Union and the United States in 1972, the Anti-Ballistic Missile Treaty forbids a national missile defense. Article I, Section II reads: "Each Party undertakes not to deploy ABM systems for a defense of the territory of its country and not to provide a base for such a defense, and not to deploy ABM systems for defense of an individual region except as provided for in Article III of this Treaty." Article III allows each side to build a defense for an individual region that contains an offensive nuclear force. In other words, the ABM Treaty prohibits our government from defending the American people, while allowing it to defend missiles to destroy other peoples.

Although legal scholars believe that this treaty no longer has legal standing, given that the Soviet Union no longer exists, it has been upheld as law by successive administrations—especially the Clinton administration—and by powerful opponents of American missile defense in the U.S. Senate.

As a side note, we now know that the Soviets violated the ABM Treaty almost immediately. Thus the Russians possess today the world's only operable missile defense system. Retired CIA Analyst William Lee, in *The ABM Treaty Charade,* describes a 9,000-interceptor system around Moscow that is capable of protecting 75 percent of the Russian population. In other words, the Russians did not share the belief of U.S. arms-control experts in the moral superiority of purposefully remaining vulnerable to missile attack.

How to stop ballistic missiles

For all the bad news about the ballistic missile threat to the U.S., there is the good news that missile defense is well within our technological capabilities. As far back as 1962, a test missile fired from the Kwajaleen Atoll was intercepted (within 500 yards) by an anti-ballistic missile launched from Vandenberg Air Force Base. The idea at the time was to use a small nuclear warhead in the upper atmosphere to destroy incoming enemy warheads. But it was deemed politically incorrect—as it is still today—to use a nuclear explosion to destroy a nuclear warhead, even if that warhead is racing toward an American city. (Again, only we seem to be squeamish in this regard: Russia's aforementioned 9,000 interceptors bear nuclear warheads.) So U.S. research since President Reagan reintroduced the idea of missile defense in 1983 has been aimed primarily at developing the means to destroy enemy missiles through direct impact or "hit-to-kill" methods.

Missile defense is well within our technological capabilities.

American missile defense research has included ground-based, sea-based and space-based interceptors, and air-based and space-based lasers. Each of these systems has undergone successful, if limited, testing. The space-based systems are especially effective since they seek to destroy enemy missiles in their first minutes of flight, known also as the boost phase. During this phase, missiles are easily detectable, have yet to deploy any so-called decoys or countermeasures, and are especially vulnerable to space-based interceptors and lasers.

The best near-term option for ballistic missile defense, recommended by former Reagan administration defense strategist Frank Gaffney, is to place a new generation of interceptors, currently in research, aboard U.S. Navy Aegis Cruisers. These ships could then provide at least some missile defense while more effective systems are built. Also under consideration is a ground-based system in the strategically important state of Alaska, at Fort Greely and Kodiak Island. This would represent another key component in a comprehensive "layered" missile defense that will include land, sea, air and space.

Arguments against missile defense

Opponents of missile defense present four basic arguments. The first is that ABM systems are technologically unrealistic, since "hitting bullets with bullets" leaves no room for error. They point to recent tests of ground-based interceptors that have had mixed results. Two things are important to note about these tests: First, many of the problems stem from the fact that the tests are being conducted under ABM Treaty restrictions on the speed of interceptors, and on their interface with satellites and radar. Second, some recent test failures involve science and technology that the U.S. perfected 30 years ago, such as rocket separation. But putting all this aside, as President Reagan's former science advisor William Graham points out, the difficulty of "hitting bullets with bullets" could be simply overcome by placing small nuclear charges on "hit-to-kill" vehicles as a "fail safe" for when they miss their targets. This would result in small nuclear explosions in space, but that is surely more acceptable than the alternative of enemy warheads detonating over American cities.

If we get into an arms race, our enemies will go broke.

The second argument against missile defense is that no enemy would dare launch a missile attack at the U.S., for fear of swift retaliation. But as the CIA pointed out two years ago—and as Secretary of Defense [Donald] Rumsfeld reiterated recently in Russia—an enemy could launch a ballistic missile from a ship off one of our coasts, scuttle the ship, and leave us wondering, as on September 11, who was responsible.

The third argument is that missile defense can't work against ship-launched missiles. But over a decade ago U.S. nuclear laboratories, with the help of scientists like Greg Canavan and Lowell Wood, conducted successful tests on space-based interceptors that could stop ballistic missiles in their boost phase from whatever location they were launched.

Finally, missile defense opponents argue that building a defense will ignite an expensive arms race. But the production cost of a space-based interceptor is roughly one to two million dollars. A constellation of 5,000 such interceptors might then cost ten billion dollars, a fraction of America's defense budget. By contrast, a single Russian SS-18 costs approximately $100 million, a North Korean Taepo Dong II missile close to $10 million, and an Iraqi Scud B missile about $2 million. In other words, if we get into an arms race, our enemies will go broke. The Soviet Union found it could not compete with us in such a race in the 1980s. Nor will the Russians or the Chinese or their proxies be able to compete today.

Time for leadership

Building a missile defense is not possible as long as the U.S. remains bound by the ABM Treaty of 1972. President Bush has said that he will give the Russian government notice of our withdrawal from that treaty when his testing program comes into conflict with it. But given the severity of the ballistic missile threat, it is cause for concern that we have not done so already.

Our greatest near-term potential attacker, Iraq, is expected to have ballistic missile capability in the next three years. Only direct military intervention will prevent it from deploying this capability before the U.S. can deploy a missile defense. This should be undertaken as soon as possible.

Our longer-term potential attackers, Russia and China, possess today the means to destroy us. We must work and hope for peaceful relations, but we must also be mindful of the possibility that they have other plans. Secretary [of State Colin] Powell has invited Russia and China to join the coalition to defeat terrorism. This is ironic, since both countries have been active supporters of the regimes that sponsor terrorism. And one wonders what they might demand in exchange. Might they ask us to delay building a missile defense? Or to renegotiate the ABM Treaty?

So far the Bush administration has not demonstrated the urgency that the ballistic missile threat warrants. It is also troublesome that the President's newly appointed director of Homeland Security, Pennsylvania Governor Tom Ridge, has consistently opposed missile defense—a fact surely noted with approval in Moscow and Beijing. On the other hand, President Bush has consistently supported missile defense, both in the 2000 campaign and since taking office, and he has the power to carry through with his promises.

Had the September 11 attack been visited by ballistic missiles, resulting in the deaths of three to six million Americans, a massive effort would have immediately been launched to build and deploy a ballistic missile defense. America, thankfully, has a window of opportunity—however narrow—to do so now, before it is too late.

Let us begin in earnest.

3

America Should Not Pursue a Ballistic Missile Defense System

Michelle Ciarrocca and William D. Hartung

Michelle Ciarrocca and William D. Hartung are scholars with the World Policy Institute at the New School for Social Research in New York City.

President George W. Bush has breathed new life into the grandiose missile defense designs of former president Ronald Reagan, whose dream of making the world safe from nuclear weapons through missile defense has not been realized despite years of research costing billions of dollars. Bush's proposals could spark a new global arms race and halt any progress on arms reductions talks. The costs of deploying a national missile defense system outweigh its benefits. The United States should instead seek security through diplomacy and by living up to treaty promises to reduce and eliminate its nuclear arsenals.

S ome dreams never die. On March 23, 1983, Ronald Reagan surprised the nation and the world by announcing an ambitious research program designed to render nuclear weapons "impotent and obsolete." Reagan acknowledged that this "formidable technical task . . . may not be accomplished before the end of this century." He was right: the U.S. has spent more than $70 billion since that time on various missile defense programs without producing a single workable device.

Under the Clinton administration, it became U.S. policy to deploy a National Missile Defense (NMD) system "as soon as technologically feasible." President Bill Clinton's commitment to missile defense was tempered by his pledge to base a deployment decision on four criteria: the overall costs of the program, the technical feasibility, an assessment of the ballistic missile threat facing the U.S., and the impact it would have on arms control and arms reduction efforts.

From "Star Wars Revisited," by Michelle Ciarrocca and William D. Hartung, *Foreign Policy in Focus*, June 2001. Copyright © 2001 by Foreign Policy in Focus. Reprinted with permission.

Although the NMD system was restructured to focus on the seemingly more realistic goal of defending all 50 states from an accidental missile launch by Russia or China or from the attack of a rogue nation such as Iran, Iraq, or North Korea, technological difficulties still abound. Given a critical test failure in July 2000 and a growing chorus of criticisms, President Clinton found himself in a safe position to delay deployment of the proposed NMD system before leaving office. In September 2000 he said, "I simply cannot conclude with the information I have today that we have enough confidence in the technology, and the operational effectiveness of the entire NMD system, to move forward to deployment." Clinton added that even if missile defenses could be made to work, they would at best add a modest margin of protection from nuclear weapons. At worst, they could spark a new, multisided nuclear arms race that would increase the risks of nuclear war.

Bush's missile defense proposals

But Reagan's dream of a shield against nuclear weapons lives on. As Frances Fitzgerald writes in *Way Out There in the Blue*, her history of what was then known as the Strategic Defense Initiative: "Every time the program seemed ready to expire, or collapse of its own weight, something would happen to bring it to life again." The latest "something" keeping the program alive is the administration of President George W. Bush.

In the rush to deploy a missile defense system, all the risks . . . surface immediately, whereas the purported security benefits . . . will not be realized for years, if at all.

With missile defense-booster Donald Rumsfeld by his side, on May 1, 2001, Bush delivered a speech at the National Defense University, sounding an awful lot like Reagan. Bush reiterated his campaign pledge calling for a missile defense system capable of defending the entire U.S., as well as "our friends and allies and deployed forces overseas," from ballistic missile attack. The Bush team proposes a layered approach that would combine the ground-based NMD system inherited from the Clinton administration with sea-, air-, and space-based components, but specifics remain vague. The administration promises that information on the cost, timing, and structure of the proposed system will be released after Secretary of Defense Rumsfeld unveils the results of his review of U.S. defense strategy.

In meetings with NATO defense ministers, Defense Secretary Donald Rumsfeld claimed that a missile defense system would "dissuade and discourage potential adversaries" while enhancing deterrence. He added, "as this program progresses, we will likely deploy test assets to provide rudimentary defense to deal with emerging threats." But allies remain skeptical. German Defense Minister Rudolf Scharping has argued that "a coherent political answer to the threats" is needed, "because technological means alone are not sufficient." Secretary of State Colin Powell fared no better in his meetings with NATO foreign ministers. They offered to "continue substantive consultations" with the U.S. but did not endorse President Bush's missile defense plan.

Unconvinced of either the missile threat or the technological merits of a missile defense system, America's allies are primarily concerned that the Bush administration will abrogate the 1972 Anti-Ballistic Missile Treaty with Russia. In a statement distributed to the NATO ministers, Rumsfeld said, "The treaty stands in the way of a 21st-century approach to deterrence." However, a hasty decision to withdraw from the treaty could seriously jeopardize future nuclear reductions in Russia's armaments and encourage China and other nations to build up their arsenals. In the rush to deploy a missile defense system, all the risks—undermining the Anti-Ballistic Missile (ABM) Treaty, sparking a new nuclear arms race, and straining U.S. relations with its NATO allies—surface immediately, whereas the purported security benefits of missile defense deployment will not be realized for years, if at all. [In December 2001, Bush announced that the United States planned to withdraw from the ABM Treaty in June 2002.]

Problems with current U.S. policy

Despite the Bush administration's determination to have a rudimentary missile defense system in place by 2004, the fact remains that none of the Pentagon's missile defense programs are up to the task, and it is *not* because the ABM Treaty is standing in the way. The annual report of the Pentagon's Director of Operational Test and Evaluation (DOT&E) outlines the daunting challenges facing U.S. missile defense programs. Assessing the ground-based NMD system, the DOT&E report warns that the system is far from ready to intercept the kinds of missiles "currently deployed by the established nuclear powers" and recommends broadening the test program to attempt to intercept real world threats that include decoys. To date, the system has failed two out of three intercept tests. A new DOT&E report claims that the one successful test used a Global Positioning System inside the mock warhead that helped guide the intercept missile to the target.

Meanwhile, the sea-based, boost-phase system, promoted by many missile defense advocates as a near-term and easy solution to the nuclear threat, is based on a missile that has yet to be designed, much less tested. The DOT&E report asserts that it is not a viable option and goes on to note that "a major upgrade to the AEGIS radar" would be required, while both the missile and kill vehicle would have to be radically redesigned. Optimistic estimates put initial deployment at 2008, with full deployment not possible before 2020.

The missile threat has been greatly exaggerated, while the consequences of deploying a NMD system have been downplayed.

The Space-Based Laser, intended to destroy a ballistic missile in its boost phase, is little more than a concept at this stage. Only a handful of components of the system have been tested, the actual testing facility hasn't even been built, and integrated flight experiments aren't expected to take place until 2010. According to a General Accounting Office report, the Air Force's new satellite surveillance package, called Space-Based Infrared System-Low, is "at a high risk of not delivering the system on time or at cost or with expected performance." The satellite network, which is to track incoming warheads and decoys, is a vital component of any expansive missile defense system. The Air Force plans to launch the network's 24 satellites in 2006, with the full system deployed by 2010. But this time frame means that the Air Force would begin deploying the satellites before adequate testing has been completed.

The continued pursuit of missile defense . . . will create a false sense of security for Americans and increase the threat of nuclear war.

The multitiered approach favored by the Bush administration will be enormously expensive, dwarfing the Congressional Budget Office estimate of $60 billion for the Clinton administration's more modest system. Estimates for the more ambitious Bush approach range from the Council for a Livable World's projection of at least $120 billion to the Center on Strategic and International Studies' (CSIS) prediction of $240 billion. In the short term, the Bush administration is planning to increase missile defense funding from the $5.3 billion allocated for FY2001 to $7.5 billion for FY2002. Under Bush, total missile defense spending could jump to $10 billion or more annually. The major contractors—Boeing, Lockheed Martin, Raytheon, and TRW—have already racked up long-term contracts for missile defense worth in excess of $20 billion, and that's *before* they reap the benefits of the new spending that will flow under President Bush's more expansive approach.

The missile threat has been greatly exaggerated, while the consequences of deploying a NMD system have been downplayed. The government's top ballistic missile analyst, Robert Walpole, has repeatedly pointed out that an attack on U.S. territory with a weapon of mass destruction has a "return address" on it, meaning the U.S. would know exactly where it came from and would launch a devastating retaliatory strike. North Korea, the supposed impetus behind U.S. missile defense efforts, is years away from developing a reliable ballistic missile that could deliver a nuclear warhead to the United States. Furthermore, Pyongyang has put its missile program on hold to pursue negotiations with Washington.

Just how big a threat missile defense could pose to U.S. security can be found in a report issued last summer by the National Intelligence Council. That report suggested that deployment of such a system would likely provoke "an unsettling series of political and military ripple effects . . . that would include a sharp buildup of strategic and medium-range nuclear missiles by China, India and Pakistan and the further spread of military technology in the Middle East."

Bush has suggested reducing the number of nuclear weapons in the U.S. arsenal to "the lowest possible number consistent with our national security" and taking these weapons off of hair-trigger alert. Bush rightly noted that, "keeping so many weapons on high alert may create unacceptable risks of accidental or unauthorized launch." But Bush also stated that "nuclear weapons still have a vital role to play in our security and that of our allies." In their more honest moments, President Bush and his advisers speak of "refashioning the balance between defense and deterrence," not replacing the cold war era "balance of terror" with a defensive shield. The seeming contradiction in the Bush view—reducing the size of the U.S. arsenal and taking forces off of high alert while provoking other nuclear powers with a massive Star Wars program—disappears if you look at the common thread uniting these proposals: nuclear unilateralism.

Spurred on by the ideological ranting of conservative think tanks like the Heritage Foundation and Frank Gaffney's Center for Security Policy, a powerful bloc within the Republican Party has increasingly come to treat negotiated arms control arrangements as obstacles to U.S. supremacy rather than as guarantors of a fragile but critical level of stability in the nuclear age. The right-wing rallying cry is "peace through strength, not peace through paper." If that means shredding two decades of international arms control agreements (most of which were negotiated by Republican presidents), then so be it.

Toward a new foreign policy

While President Bush and his advisers are trying to gather international support for their dubious and vague missile defense proposal, they're squandering valuable time that could be used to promote cooperative measures for the reduction of both U.S. and Russian nuclear arsenals. A series of provocative actions toward Russia—from the expansion of NATO to references to Russia as a "top proliferator" of ballistic missile technology to persistent statements about the U.S. withdrawing from the ABM Treaty—has stalled momentum toward U.S.-Russian nuclear weapons reductions.

President Bush has tied deployment of a missile defense system to deep reductions in the U.S. nuclear arsenal in an attempt to allay Russian and international opposition. Bush has also hinted at a sort of "grand compromise" to gain Russia's approval for amending the ABM Treaty to allow for a missile defense system. The proposed package deal could include military aid, joint antimissile exercises, and arms purchases for Moscow. But the offer seems more salesmanship than substance, with no genuine attempt to ease Russia's fundamental concerns. Meanwhile, Russian President Vladimir Putin has warned that U.S. violation of the ABM Treaty would force Russia to augment its nuclear capability by mounting multiple warheads on its missiles. At the same time, Putin suggested that both the START I and START II treaties would be negated by the U.S. abrogating the ABM Treaty. The termination of these treaties would also eliminate verification and inspection requirements and allow Russia to hide its nuclear capabilities.

Taking innovative steps to get nuclear reductions back on track should be the top priority of U.S. policymakers, and such reductions should not be tied to a U.S. missile defense system. Nuclear arms reductions between the U.S. and Russia have been stalled since the signing of the START II Treaty in 1993. The treaty, which would reduce each nation's nuclear arsenals to 3,500, has yet to be ratified by the U.S. Senate. President Putin has suggested cutting even further, to 1,000 or 1,500 nukes each, while President Bush has voiced a similar position but avoided an exact number.

Additionally, instead of cutting more than $72 million in funds intended to help safeguard and dispose of Russian nuclear material, President Bush should be showing his commitment to nonproliferation by increasing the budget for these activities. A bipartisan commission issued a report in January 2001 calling the risk of theft of Russian nuclear materials "the most urgent unmet national security threat" facing the U.S. and urged sharp increases in spending for the Russian programs. Getting U.S.-Russian nuclear reductions back on track and supporting multilateral efforts toward nuclear abolition would also give the U.S. much greater credibility in promoting wide-ranging security discussions between India and Pakistan aimed at capping and eventually eliminating their nascent nuclear programs.

As for North Korea, Iran, or Iraq, there are other methods of dealing with the threat of a ballistic missile attack from these nations that would be far less costly and far more effective than building a multibillion-dollar missile shield. But instead of picking up where the Clinton administration left off in talks with Pyongyang, Bush started his term by delaying further negotiations until his administration could conduct a comprehensive review of U.S. policy toward North Korea. As Spurgeon Keeny, president of the Arms Control Association, notes, Bush's actions (or lack thereof) are "widely perceived internationally as intended to preserve, and even enhance, the North Korean ballistic missile threat so that it can serve as the rationale for early deployment of a national missile defense." Initial Bush administration efforts to restart the talks with North Korea aroused skepticism when new demands were laid on Pyongyang in the area of conventional force reductions without indicating when or whether Washington would meet its original obligations under the framework agreement. If implemented as planned, the framework agreement could scale back and eventually eliminate Pyongyang's nuclear weapons and ballistic missile programs as part of an overall improvement in U.S.-North Korean economic and political relations. President Bush should fulfill America's long-overdue commitments under the nuclear framework agreement with North Korea and should continue to support South Korea's efforts at cooperation and reconciliation with North Korea.

Ultimately the U.S. and other nuclear powers should strive for a nuclear-weapons-free world by living up to their commitments, signed 30 years ago, under Article VI of the Nuclear Non-Proliferation Treaty (NPT) "to reduce and eventually eliminate their vast arsenals of nuclear weaponry." On May 20, 2000, at the conclusion of the sixth review of the NPT, the U.S. and 186 other countries came to a global consensus on nuclear disarmament, declaring it the "only absolute guarantee against the use or threat of use of nuclear weapons." The U.S. must lead the way toward this goal.

The continued pursuit of missile defense will have far-reaching consequences for the future of arms control and the goal of nuclear abolition. It will create a false sense of security for Americans and increase the threat of nuclear war for the world. A modest missile defense program of research, in the range of a few hundred million dollars per year focused on primarily improving the performance of a medium-range defensive shield to replace the current Patriot system, is justified as a way to limit the potential damage posed by the use (or threat of use) of medium-range missiles. But the main focus of Washington's energy and resources should be on preventive measures, which are far more effective at reducing the threat of nuclear war than any pie-in-the-sky defensive schemes.

4

America Should Pursue a Limited Missile Defense System

Charles V. Peña and Ivan Eland

Charles V. Peña and Ivan Eland are defense policy experts at the Cato Institute, a libertarian public policy research foundation.

The debate over missile defense has been distorted by rhetorical excesses from both proponents and opponents. Opponents of missile defense should recognize that a threat to the United States does exist and that a national missile defense system can enhance America's security. However, such a system should be land-based and designed solely to protect America from a limited missile attack. More grandiose space-based missile defense systems touted by some proponents should be rejected as being unnecessary and potentially dangerous.

To date, the debate surrounding national missile defense (NMD) has been dominated by political rhetoric. Supporters (usually conservatives) often paint a "doom-and-gloom" picture, pointing out that the United States is vulnerable to attack by ballistic missiles. Critics (usually liberals) defend the Anti-Ballistic Missile (ABM) Treaty as the cornerstone of nuclear deterrence and stability and argue that any defensive deployment would upset the balance between the offensive strategic forces of the United States and Russia.

Opponents of NMD, who use preserving the ABM Treaty as an argument to forestall deployment of a defense, need to acknowledge that the threat of attack by long-range ballistic missiles from "rogue" states may become real. Opponents also need to realize that the United States can build a limited NMD without disrupting the bilateral strategic nuclear balance. Supporters of NMD need to acknowledge that NMD is not a panacea for the full spectrum of threats from rogue states—that long-range ballistic missiles are only one of the options available to those states. Supporters also need to recognize the daunting technological chal-

Excerpted from "Strategic Nuclear Forces and Missile Defense," by Charles V. Peña and Ivan Eland, *Cato Handbook for Congress: Policy Recommendations for the 107th Congress* (Washington, DC: Cato Institute, 2001). Copyright © 2001 by the Cato Institute. Reprinted with permission.

lenge that NMD poses and not seek to rush its development.

A limited NMD, which would afford the United States protection against long-range ballistic missile threats from rogue states, seems feasible and probably can be deployed at a reasonable cost. The elements of the limited land-based system under development during the Clinton administration can provide such capability.

No matter what the threat, however, the development of an NMD system should proceed at a measured pace because an excessively rapid development program could waste taxpayer dollars on an ineffective system. NMD should remain a research and development (R&D) program until it has been thoroughly tested under realistic operational conditions. Only then should a decision be made about its deployment.

Any defense expenditures—including those on missile defense—must be commensurate with the threat. More robust missile defenses are not justified by the present limited threat. Also, sinking large amounts of money into more comprehensive missile defenses—when even the limited land-based system might fail because of technical problems or lack of adequate testing—is questionable.

A limited NMD is needed for a limited threat

Although it is not certain that North Korea will be capable of launching a missile attack against the United States by 2005, the R&D program for NMD is being rushed to have a system deployed by that date. Even if the threat from North Korea did materialize by that date, the United States would probably be able to use its offensive nuclear force to deter a missile attack from North Korea, another "rogue" state, or any other state. Thus, NMD would be a backup system against a missile attack from a pariah state. Rushing development increases the probability that the system will ultimately be delayed, will experience escalating costs, or will simply not work.

More important, rogue states have or will have options for striking the United States other than long-range ballistic missiles. Such countries already possess short- and medium-range ballistic missiles that could be launched from ships operating in international waters off the U.S. coasts. They also may possess or could acquire cruise missiles that could be launched from ships or, possibly, aircraft. Finally, terrorist attacks using weapons of mass destruction are an option readily available to rogue states (or groups they sponsor), especially given the open nature of American society.

> *NMD [national missile defense] is not a panacea for the full spectrum of threats from rogue states.*

Such threats to the American homeland may be more inexpensive, accurate, reliable, and thus more probable than that posed by intercontinental ballistic missiles (ICBMs) launched from rogue states. Even the most hostile pariah state is likely to hesitate to launch from its territory an ICBM against the United States. U.S. satellites can detect the origin of such long-range missile launches, and the world's most powerful nuclear force would almost certainly retaliate against the attacking nation. In contrast, the origin of terrorist attacks or missile launches from ships or aircraft may be harder to determine, which makes U.S. retaliation—and therefore deterrence—more difficult. The existence of the other threats does not, of course, refute the argument that long-range ballistic missiles also pose a threat and that the U.S. government should combat the threats that can be defeated. But we must understand that long-range ballistic missiles will be just one of several possible threats.

Not all proposals for deploying a "national" missile defense system live up to their name.

None of the proposed NMD systems will have a defensive capability against either short-range ballistic missiles or cruise missiles—delivery systems for which rogue states and others may already possess. The best reason to have a limited missile defense may be the possibility of accidental—rather than intentional—launches from such states and limited accidental launches from established nuclear powers. Pariah states with newly acquired long-range missiles and nuclear warheads may have poor early warning systems, only rudimentary command and control over such forces, nonexistent nuclear doctrine, and insufficient safeguards against an accidental launch. In addition, in the past, Russia's decrepit early warning systems have almost led to accidental launches.

Nevertheless, the primary threat from accidental or intentional launches from rogue states is likely to be relatively modest (a few ICBMs) and unsophisticated (their missiles are unlikely to have multiple warheads or sophisticated decoys), requiring an equally modest response. A limited ground-based NMD system of 100 or so interceptors could provide sufficient defensive capability against such threats.

The limited threat does not warrant "international" defenses

Not all proposals for deploying a "national" missile defense system live up to their name. Many are for "international" missile defense systems that would also defend U.S. allies and "friends," even though they are wealthy enough to build their own missile defenses. For example, some policymakers and analysts on both the left and the right advocate sea-based missile defense as a substitute for the limited land-based system, which is designed to protect only the territory of the United States. Many conservatives would like to build a more comprehensive and robust layered defense consisting of sea- and space-based weapons or land-, sea-, and space-based weapons.

Proponents of sea-based NMD argue that such a system can be deployed more quickly and will be less expensive than the limited land-based system. Some argue that the Navy Theater Wide system (a system that is currently being designed to provide midcourse intercept capability against slower, shorter-range theater ballistic missiles) can be upgraded to destroy long-range ICBMs in their boost phase (when under powered flight at the beginning of their trajectories). To intercept faster, longer-range missiles in the boost phase, a new, faster interceptor would need to be developed. That interceptor would probably not be compatible with the vertical launchers of Navy ships. Forward-deployed sea-based NMD might also experience operational difficulties, including greater vulnerability to attack, and detract from the Navy's other missions.

Even if a sea-based missile defense could be developed faster and more inexpensively than the more mature land-based system (a dubious proposition since the sea-based system would depend on sensor, communication, and kill vehicle technology being developed for the land-based system), critical gaps in coverage would necessitate supplementing the sea-based system with expensive space-based weapons. Unlike land-based NMD, sea-based NMD is not a stand-alone system.

The main objective of observers who support more comprehensive, robust, and layered missile defense systems does not seem to be defense of the U.S. homeland. Instead, their aim seems to be to create a stronger shield behind which the United States can intervene against potential regional adversaries possessing weapons of mass destruction and the long-range missiles to deliver them. According to that reasoning, if such adversaries cannot threaten the United States or its allies with catastrophic retaliation, U.S. policymakers will feel more confident in intervening militarily. But because no missile defense system can guarantee that all incoming warheads will be destroyed, such an increase in U.S. military activism could actually undermine U.S. security in a catastrophic way. Thus, development of a missile shield should be confined to the more limited land-based "national" system under development during the Clinton administration.

Deep reductions in offensive nuclear forces combined with the deployment of a limited land-based NMD system would greatly enhance U.S. security.

Also, many advocates of sea- and space-based weapons want to protect U.S. friends and allies. But the United States should refuse to cover those wealthy nations—which spend too little on their own defense and already benefit from significant U.S. security guarantees—with a missile shield. A layered international missile defense that adds sea- and space-based weapons will escalate the costs of an NMD system dramatically. In addition, an international defense is not warranted by the limited threat and should not be used to defend rich allies who can afford to build their own missile defenses.

A limited land-based NMD (for example, a hundred or more ground-based interceptors designed to defend against tens of warheads from rogue states) would not enable the United States to undermine nuclear stability by threatening Russia's surviving offensive nuclear forces (even at reduced levels, numbering in the hundreds or thousands of warheads), but more robust defenses might do so. In addition, deploying robust defenses might cause an "action-reaction" cycle with China. As China modernizes and builds up its small nuclear forces (which will probably happen whether or not U.S. defenses are deployed), robust defenses are much more likely to cause a larger Chinese buildup than is a limited NMD. If Congress encouraged the new administration to pair a limited missile defense with deep cuts in the U.S. offensive nuclear arsenal, the United States could send a signal to both powers that it was not trying to achieve strategic advantage.

Combine limited NMD with deep cuts in offensive strategic weapons

The most prudent course of action is to pursue development of a limited NMD system to defend against rogue state threats and accidental launches, renegotiate the ABM Treaty with the Russians, and continue further strategic arms control negotiations under the Strategic Arms Reduction Treaty (START) process. In fact, the Russians have intimated that they might be willing to accept changes to the ABM Treaty to allow a limited NMD in exchange for even deeper cuts in strategic offensive forces. . . .

Lower numbers of warheads in the inventories of Russia and the United States would probably mean lower numbers of warheads on alert status, and lower numbers of warheads on alert status would substantially reduce the risk of an accidental nuclear launch. The lower inventory levels would also mean that fewer nuclear warheads would be available to be stolen or sold to rogue states (that possibility is a particular concern for the aging and insecure Russian nuclear stockpile).

The United States should pursue a limited land-based NMD system to defend against accidental launches or intentional missile attacks from rogue states and simultaneously renegotiate the ABM Treaty with the Russians. That conclusion does not imply that the ABM Treaty is sacrosanct or the cornerstone of strategic stability. Rather, it simply acknowledges that concerns about stability and deterrence vis-à-vis Russia are legitimate and cannot be ignored. To simply ignore the ABM Treaty and Russian concerns would needlessly antagonize Russia at an inopportune time (much as the United States did by expanding NATO and conducting the war in Kosovo)—potentially throwing away the gains of START II, START III, and other arms control agreements. Of course, any renegotiation would have to retain the basic aim of the ABM Treaty—restraining defenses so that neither the U.S. nor the Russian strategic arsenal would be undermined—while permitting limited land-based NMD against rogue states.

Working with the Russians to renegotiate the ABM Treaty—rather than unilaterally withdrawing from it—has advantages. In response to a unilateral U.S. withdrawal from the treaty, the Russians could sell rogue states the countermeasures (for example, decoys) to defeat any NMD system, refuse to help stem the spread of Russian weapons of mass destruction to such states, or maintain large numbers of nuclear weapons on alert.

It is possible to achieve a "balance" between strategic offensive arms control, the ABM Treaty, and NMD against the emerging threats. But that balance will not be achieved without dispensing with the overheated political rhetoric on both sides of the issue. Deep reductions in offensive nuclear forces combined with the deployment of a limited land-based NMD system would greatly enhance U.S. security.

5

America Should Pursue a National Defense Against Cruise Missiles

Michael O'Hanlon

Michael O'Hanlon is a foreign policy scholar at the Brookings Institution and the author of Technological Change and the Future of Warfare.

Cruise missiles (jet-powered low-flying guided missiles) can pose a serious threat to the national security of the United States, especially if terrorist groups gain possession of them. Such missiles are relatively small, can be launched from ships, are hard to detect, and could be configured to carry chemical or biological weapons. However, cruise missile attacks could possibly be defended against by using radar to detect them and interceptor missiles launched from ships, fighter aircraft, or surface-to-air missile sites to destroy them. The United States should seriously consider establishing a national defense against cruise missile attacks and should increase research funding in this area.

The September 11, 2001 terrorist attacks have reshaped whole swaths of debate over U.S. foreign and national security policies. Certainly, the issue of homeland security is a case in point. In that context, it was inevitable that the various partisans and detractors of national missile defense, and those with contending views of how homeland security should be organized, would use the September 11 tragedy as evidence for their particular position. And they have. For example, those who have held the very idea of national missile defense to be a form of inanity—if not insanity—have argued that since no imaginable deployment of national missile defenses could have prevented the September 11 tragedy, this proves how bad an idea it is. This is a little like arguing that if a person has purchased homeowner's insurance, he or she has no need for auto or life or medical insurance. The dangers to the security of a nation are multiple, no less than the dangers to the security of individuals.

That said, it does not follow that all kinds of insurance policies are equally necessary or that all insurance products are equally wise and cost-effective investments. All such policies and products require study, for nations no less than for individuals. Many such studies are going forward. But while there are ongoing and fairly well-rehearsed debates afoot over homeland security organization and national missile defense, one area that has slipped through the cracks of public consciousness concerns defense against cruise missiles. The analysis that follows represents a study of the technical requirements and costs of a defense against cruise missiles.

A national cruise missile defense

The United States today utterly lacks an effective cruise missile defense plan. But apart from the obvious post-September 11 concern about the hijacking of domestic flights, defending against cruise missiles has probably become the most challenging air defense problem for the United States in this era. . . .

Cruise missiles are prevalent around the world, with 75 countries owning a total of about 75,000. Most at present are antiship missiles, but a number of countries are working on converting some to land-attack variants. Cruise missiles are small and relatively easy to hide on ships or other vehicles that could approach U.S. territory before the missile was fired. They are hard to detect when launched or even as they approach their targets, not only because of their small size but also because of their modest infrared heat signature (especially by comparison with ballistic missiles) and their ability to fly low, using terrain for cover.

The United States today utterly lacks an effective cruise missile defense plan.

Indeed, cruise missiles are small and inexpensive enough that it may not be beyond the means of terrorists to acquire them. Reconfiguring a standard cruise missile to carry a primitive nuclear warhead, likely to weigh half a ton or more, is probably beyond the abilities of terrorists, but outfitting a cruise missile with a dispensing mechanism for distributing chemical or biological agents or radiological materials may be feasible. In this sense, the cruise missile threat to the United States should be construed as one that could be posed by terrorists as well as other states seeking a means of coercion or deterrence.

To reliably protect the country against cruise missiles is admittedly a very difficult proposition, given the multiplicity of possible launch points, approach trajectories and targets. But a system of radars, perhaps held up by aerostat balloons, together with the existing network of U.S.-based fighter aircraft and a new series of surface-to-air missile sites, could provide at least some coverage of all of the nation's borders. That network might not provide leakproof defense in all cases, but it could stop most small attacks with high confidence and deny any attacker certainty that his cruise missiles would reach U.S. territory once fired.

Technologies are being developed that could perform some of these tasks. But the pace of research is too slow, with a target date of roughly

2010 for completing a master plan on cruise missile defense for overseas battlefields. Cruise missile defense research efforts should grow to at least the cost of individual Theater Missile Defense (TMD) programs—$200 to $300 million a year above current levels.

What would be involved in eventual deployment of a cruise missile defense, something that, owing to the nature of the threat, might have to be attempted in the course of this decade on roughly the same time frame as deployment of ballistic-missile defense? Given that the United States typically spent well over $10 billion a year on air defense during the early decades of the Cold War, when it worried seriously about the Soviet bomber force, the scale of the necessary effort could be substantial. Large numbers of radars as well as widely-distributed interceptor missile bases would likely be needed to defend the vast perimeter of the United States, including non-continental states and possessions.

What a national cruise missile defense could look like

More specifically, the concept of operations for national cruise missile defense might look roughly like this. As an outermost layer of defense, Navy ships might intercept foreign ships known to have no legitimate business anywhere near American waters or the U.S. exclusive economic zone (which extends 200 nautical miles beyond American terra firma). However, this approach would likely work only against the vessels of certain foreign navies; it would risk violating international laws of shipping if employed wantonly beyond U.S. coastal waters (which extend out only twelve nautical miles from the coast). As a second and better layer, the Coast Guard could monitor ship traffic, using its new requirement for 96-hour notice of port visits to find commercial ships with suspicious owners, crews or cargoes. Its cutters could intercept and board such ships, generally tens of miles from land.

> *Outfitting a cruise missile with a dispensing mechanism for distributing chemical or biological agents . . . may be feasible.*

Clearly, even if they realized their maximum potential, such Coast Guard efforts could fail to prevent the launchings of missiles from dozens of miles off American shores. Both the Coast Guard and the Air Force would also be hard-pressed to prevent the launchings of cruise missiles from commercial aircraft secretly outfitted for such purposes. For these reasons, actual defenses against cruise missiles would be needed, too.

To provide such defenses, the United States would need to see the missiles coming, recognize them as cruise missiles and not small aircraft, and then launch nearby interceptors quickly enough to destroy the missiles before they could reach American shores. Accomplishing the detection mission requires continuous coverage of all approaches to U.S. territory by a system of radars based on land, ships, aircraft or balloon aerostats. Accomplishing the intercept mission, by contrast, requires one to estimate the distance at which enemy missiles might be launched.

Clearly, if ships or planes could launch missiles arbitrarily close to U.S. territory, even a very dense network of shore-based interceptors could prove insufficient. Enemy forces could approach to within say a couple miles of their target, then fire their missile or missiles; unless interceptors were located within a few miles of any such point, they could not reach the enemy missile quickly enough to prevent it from striking its target.

So how much warning would likely be available, and from how close could enemy missiles realistically be launched? Assuming that the main threat of cruise missile attack would arise from ships, the problem may be solvable. If Coast Guard monitoring can work effectively, suspicious ships entering U.S. territorial waters and contiguous zones should certainly be identifiable. Assuming that they could be stopped at the outer edges of contiguous zones—and fired upon immediately if they did not—it should be possible to prevent the launch of cruise missiles from closer than roughly twenty nautical miles or about 25 standard miles. For a relatively simple subsonic cruise missile, that might correspond to five minutes of flight time in which interceptors could do their work before targets on land were struck.

These timelines translate into a difficult but not impossible job of protecting coastal regions. If defenses had to be deployed uniformly around American coastlines, and interceptor missiles could accelerate quickly and then travel at roughly one kilometer per second, it might suffice to have a base of several interceptors every fifty miles or so. If, however, the main goal was to protect larger towns and cities as well as key ports and infrastructure, fewer interceptors would be required along certain stretches of coastline, where spacing might be every 100 to 200 miles. Either way, given that the length of U.S. coastal zones is several thousand miles (even after one "smooths out" complex coastlines and waterways, and focuses just on the main perimeter of the country), many dozens of interceptor bases would be needed.

What about radar coverage? Here the key variables are the altitude at which the radars can be situated, together with their power and range. Radars on the ground, even if on hills, cannot see very far due to the curvature of the earth and the low altitudes at which cruise missiles customarily fly. If cruise missiles could fly as low as fifty feet, and radars often could not be located near shorelines on hills higher than 100 to 200 feet (for illustrative purposes), radars would be needed every fifteen to twenty miles all around the perimeter of the United States. By contrast, if airplanes or aerostats at several thousand feet altitude could be used to provide surveillance and targeting information, they could be spaced every 100 to 200 miles, reducing the total need to several dozen for the entire country. However, two or more such radars would probably be needed to keep one platform on continuous station.

Total costs

What total costs are implied by such a cruise-missile defense architecture? Short-range missiles would typically cost perhaps $1 million to $3 million apiece; their associated ground installations would add to that tab. If ten were based at each of 100 locations, total acquisition costs for the missiles and their ground support might be $5 billion to $10 billion. Radars based in the air might cost anywhere from several tens of millions to several hundred million each, depending on their sophistication and on how successful engineers may be at finding inexpensive solutions. If 100 were needed at $100 million each, costs would be $10 billion. If, by contrast, costs could be held to those for advanced unmanned aerial vehicles such as the Global Hawk, expenses might be half as much or less. In addition, given the long endurance of that aircraft, fewer total aircraft might be needed to keep a certain number in the sky at a time, so total acquisition costs for the airborne platforms might be held to $3 or $4 billion.

Operating costs would of course be additional. For an unmanned aerial vehicle (UAV), annual operating costs might be $3 million to $5 million. For a medium-sized aircraft, costs would more likely be $5 million to $10 million each year. Assuming roughly 100 such aircraft, annual costs would thus be $300 million to $1 billion, depending on the capabilities of the airframe and the radar.

All told, a rudimentary cruise missile defense for the United States could probably cost $10 billion to $20 billion to deploy. It would cost roughly as much to operate over a twenty-year period. Overall, these figures would represent a modest investment on the scale of national ballistic missile defense, but a large number relative to most other homeland security requirements.

These numbers ignore several other factors that could change costs significantly. However, two of them would tend to raise costs, and the others would tend to lower them, so the net effect would likely be quite modest, and the above estimates might turn out to be reasonably accurate.

One factor that would raise costs would be the need for command, control and battle management infrastructure to connect radars to each other and to interceptors. As one guidepost to possible costs, command and communications facilities for the proposed midcourse ballistic missile defense system designed by the Clinton Administration are expected to cost just over $2 billion (or about 10 percent of the total). In addition, costs of $1.5 billion were expected for construction of major sites; similar demands would arise for the cruise missile defense discussed above.

Lower costs would likely result from the fact that existing Federal Aviation Administration radars might make it unnecessary to purchase and deploy radars in some parts of U.S. coastal regions. In addition, as noted above, aerostat balloons with very long endurance might be able to replace ground radars in a number of locations at considerably lower cost.

This analysis shows that a defense against ship-launched cruise missiles is desirable and achievable, but difficult and expensive. As with all other defense systems, any cruise missile defense effort would have to compete with other programs of roughly equal importance. If we rule out, as we should, both technological impossibility and technological inevitability arguments, and if we recognize that resources for defense are far more elastic in a national crisis than almost anyone thinks they are in normal times, then the question of cruise missile defense falls into the familiar and proper context of political judgments about competing needs. Such judgment, however, needs to be exercised. The United States should seriously consider the desirability and feasibility of a national defense against cruise missiles armed with chemical, biological or radiological payloads.

6

The September 11 Terrorist Attacks Strengthen the Case for Missile Defense

Baker Spring

Baker Spring is a research fellow for the Heritage Foundation, a conservative public policy research institute.

The September 11, 2001, terrorist attacks on New York and Washington, by proving that terrorists will use any means to kill Americans, bolster the case for missile defense. If terrorists gain possession of missiles, the results could be even more horrific than the September 11 attacks. Americans do not have to choose between additional counterterrorism measures and missile defense—both are essential to homeland security.

Following the September 11, 2001, terrorist attacks on New York and Washington, Members of Congress unequivocally gave their support to the President to expose those who supported the terrorists and hold them accountable. Yet barely one week later, some opponents of missile defense began to use the tragedy to support their case that the United States does not need a missile defense system. For example, Representative John F. Tierney (D-MA) stated that "This type of incident . . . is much higher on the list of threats than anything the president would address with his national missile defense program." Such observations are grossly misleading.

The horrific events of the past week have proven beyond any doubt that terrorists will use any means, at any cost, to devastate America. With the proliferation of ballistic missiles to rogue states like North Korea, the likelihood that terrorists and despots will use these weapons of mass destruction to attack U.S. territory has grown substantially.

In reality, the terrorist attacks on New York and Washington bolster the case for fielding missile defenses sooner, rather than later, to protect Americans. The reasons:

   • *Defending against both terrorism and missile attack is the government's moral obligation.* Both forms of attack obviously represent serious threats to national security. The U.S. military cannot ignore other threats to security now that it is faced with a serious terrorist threat at home.

Moreover, defending Americans is not an either/or proposition. Protecting the people, territory, and institutions of the United States is the government's first and most critical responsibility. As the Rumsfeld Commission [a bipartisan commission chaired by Donald Rumsfeld established by Congress to assess the ballistic missile threat] delineated in its 1998 report to Congress, the threat of missile attack is clear and growing (just as the threat of terrorist attack looms ever larger). The commission warned that states such as North Korea "would be able to inflict major damage on the U.S. within about five years of a decision to acquire such a capability." Shortly after that report was released, North Korea surprised the intelligence and defense communities by launching a rocket over Japan. Many now predict that its second-generation rocket is nearing completion; with a range of more than 6,000 miles, it could reach the western United States. It is indeed frightening to think of such a capability falling into the hands of terrorists.

The terrorist attacks on New York and Washington bolster the case for fielding missile defenses sooner, rather than later.

Critics who would use [the September 11] terrorist attacks to force Congress to decide to fund either counterterrorism *or* missile defense should simply be ignored. Their logic is disingenuous; in the medical world, it would lead doctors to the absurd conclusion that having vaccinated someone against polio, there is no need to worry about mumps or measles. The United States needs a balanced national security policy that addresses the full array of threats to American lives, including the expanding threat posed by ballistic missiles.

Homeland defense requires missile defense

• *Homeland defense is insufficient without missile defense.* Members of Congress, faced with the fact that the terrorists used civilian airplanes as suicide bombs, have rallied behind the cause of increasing homeland defense. This is appropriate, but a comprehensive policy for defending the homeland must provide for defenses against *all* forms of attack, whether from the air or sea, cruise missiles, ground troops, terrorists, or ballistic missiles.

The United States spent more than $10 billion to counter terrorism in the past year, but this did not prevent the attack on September 11. Congress and the American people now see the need to spend more to improve intelligence-gathering capabilities, airport security, and defensive systems to repel any attack on U.S. soil. Moreover, they recognize that the new breed of terrorists are seeking massive "collateral damage" in terms of loss of life and devastation.

If increased security on the airlines deters terrorists from using airplanes, the likelihood grows that tomorrow's terrorists will want to use missiles to wreak destruction. Consider Muammar Qadhafi's chilling words to his followers after the U.S. military had responded to his terrorists' bombing of a Berlin discotheque in 1986: "If we had possessed a deterrent—missiles that could reach New York—we would have hit it at the same moment."

• *While systems are in place to thwart terrorism, the nation still has no defense against missile attack.* Airport security measures and other systems were in place before the September 11 attack, and the intelligence community has had the authority to preempt terrorist actions. Regrettably, these were not sufficient to prevent the attacks. More must be done, and Congress is right to authorize immediate funding to increase counterterrorism.

Terrorist groups, not just states, have the means to buy ballistic missiles.

Today, however, there are no defenses against missile attack—and America's enemies know this. Arms control agreements and military cutbacks in missile defense programs not only delay progress, but also undermine national security. As the Rumsfeld Commission report makes clear, waiting even five years to fund missile defense programs is unwise and unethical, and flies in the face of the U.S. Constitution's mandate to provide for the common defense and Congress's mandate to deploy a national missile defense in the 1999 National Missile Defense Act (P.L. 106-38).

The missile threat

• *Missile attacks will be far more destructive than the September 11 assaults.* The world has been justifiably horrified at the loss of so many innocent lives during the terrorist attacks in New York, Washington, and Pennsylvania. This assault has proven that terrorists are escalating their tactics. From killing and injuring a limited number of people in order to frighten a large number, they seem determined to kill and injure large numbers of people to spread fear and demoralize nations.

The National Commission on Terrorism reported in 1998 that "Now, a growing percentage of terrorist attacks are designed to kill as many people as possible." A ballistic missile carrying nuclear, chemical, or biological weapons fits this pattern because it would result in deaths and injuries many times greater than in the devastation of September 11. A 1979 Office of Technology Assessment study estimated that, if two one-megaton nuclear warheads were to strike Philadelphia, the explosion alone would kill 400,000 people. Larger threats mandate more coherent and dedicated defenses, not less.

• *Terrorist groups, not just states, have the means to buy ballistic missiles.* Some may be tempted to assume that only states, not terrorist groups, have the means to field and launch ballistic missiles. For certain categories of large, long-range missiles, this may be true. Given the significant trade in smaller, shorter-range missiles, however, well-funded and organized terrorist groups, and especially those benefiting from state sponsorship, have the capacity both to obtain and launch such missiles.

Even these shorter-range missiles could reach U.S. territory. The 1998 Rumsfeld Commission report, for example, cited the threat to U.S. territory posed by short-range ballistic missiles that are launched from ships sitting in international waters off the U.S. coasts. Recent evidence shows that billionaire Osama bin Laden has funded attacks on U.S. embassies in Africa and the destroyer USS *Cole* in Yemen.

A larger issue also must be considered. As the National Commission on Terrorism pointed out, "Five of the seven nations the United States identifies as state sponsors of terrorism have programs to develop weapons of mass destruction." The means to deliver these nuclear, chemical, and biological weapons are ballistic missiles. States that would knowingly provide those agents to a terrorist group during times of chaos could well lose "control of the terrorists' activities."

A shield from retaliation

• *Missile defenses are needed to shield the United States from retaliation should it take action against terrorist-harboring states.* Nation-states still maintain the largest inventories of ballistic missiles and pose the most serious threat of missile attack against U.S. territory, U.S. forces overseas, and U.S. allies. Following the September 11 attack, the United States made it clear that it will also hold accountable any state that harbors and supports the terrorists responsible for this act. If such states have ballistic missiles in their arsenals, they conceivably could retaliate against a unilateral or multilateral action by launching a missile attack on U.S. territory, especially if the "United States were distracted by a major conflict in another area of the world." Military prudence dictates that the United States have a missile defense capability in place to defend America.

• *The Cold War policy of "mutual assured destruction" embodied in arms control treaties is not sufficient to deter terrorist missile attacks.* Cold War agreements such as the 1972 Anti-Ballistic Missile (ABM) Treaty limit the ability of the United States to develop and deploy effective missile defenses. Proponents of arms control often argue that the United States will not face a missile attack in the first place because missiles "carry a return address" as well as the threat of overwhelming U.S. retaliation in kind. The ABM Treaty makes a virtue of U.S. vulnerability based on the belief that nations that value human life will not use their nuclear weapons if there is retaliation in kind. But this principle embodied in arms control treaties will not deter terrorists from acting: Today's suicide terrorists demonstrated on September 11 that they do not value human life.

In light of the current circumstances, the principle of U.S. vulnerability is discredited and is becoming exceedingly unpopular. President Bush is right to tell the world that it is time to move beyond outdated paradigms to forge a new security environment—one not threatened by terrorists or the proliferation of weapons of mass destruction.

• *Nuclear retaliation is not appropriate for every kind of attack against America.* Some opponents of missile defense believe that the United States has an effective nuclear deterrent that, if necessary, could be used to respond to attacks on the homeland. But no responsible U.S. official is suggesting that the United States consider the use of nuclear weapons in response to the horrific September 11 attacks. In most cases of attack on the

United States, the nuclear option would not be appropriate, but a defense response will almost always be appropriate. The United States needs to be able to resort to defensive options.

• *The United States simply cannot afford not to address both the missile threat and terrorism.* Some critics argue that since there are only finite resources for defense, the nation must choose which threats to address. This is grossly naive. The government simply cannot continue to wait until after a missile attack occurs to fund missile defense programs. The new consensus in Washington and across the country following the September 11 attacks provides policymakers and the American public an opportunity to stop merely reacting to terrorist attacks and start constructing a military force that addresses all threats to the United States. If the people of America really want adequate defenses, they will pay for them. Congress should make sure that all national security and defense programs are fully funded.

Defending America

The argument that the September 11 terrorist attack disproves the immediacy of the missile threat to America is not just wrong; it is misleading. There is no choice for Washington when it comes to defending Americans. Both terrorism and missile attack are growing threats to national security, and all such threats deserve dedicated, systematic, and comprehensive responses.

7

The September 11 Terrorist Attacks Weaken the Case for Missile Defense

Michael T. Klare

Michael T. Klare is a professor of peace and world security studies at Hampshire College in Amherst, Massachusetts. His books include Rogue States and Nuclear Outlaws.

Making missile defense a leading national priority seems questionable after the tragic events of September 11, 2001. A ballistic missile defense system would not have prevented the terrorist attacks in New York and Washington. America should put its efforts to construct an expensive and perhaps unworkable missile defense system on hold and instead concentrate on preventing future terrorist attacks.

Almost every aspect of U.S. military policy is likely to be affected by Tuesday's [September 11, 2001] terrorist attacks in New York and Washington, but one that is certain to come under intense scrutiny is the Bush administration's plan for a national missile defense (NMD).

In particular, President George W. Bush's claim that NMD represents the single most important priority for U.S. "homeland" defense is bound to appear highly dubious in light of the apparent success of a . . . terrorist network in causing massive damage to major U.S. institutions and facilities.

As envisioned by the White House, NMD will protect the U.S. against ballistic missile attacks launched by "rogue" states like Iran and North Korea. No such nation currently possesses operational missiles with the range to strike the United States, but the administration claims that one or more of them might acquire such a capability in the next five to 10 years. To defeat such attacks, the Pentagon proposes to spend hundreds of millions of dollars on anti-missile defense systems.

Until Tuesday [before the terrorist attacks], the debate over NMD has largely revolved around two questions: First, is the "rogue state" missile threat so severe as to justify the expenditure of hundreds of millions of dollars on missile defenses? Second, can such defenses actually be made to work?

Although the White House insists that the answer to both questions is "yes," many in Congress and elsewhere have expressed skepticism about these assurances.

A new question

Now a new question will emerge: Is NMD really needed when the United States faces a very genuine threat from unconventional forces using improvised weaponry: civilian American aircraft hijacked at American airports? Even if NMD were fully functional and all of its systems worked flawlessly, it would be of no use in defending against attacks of this type.

Given the evident skill and ingenuity of those responsible for Tuesday's attacks, it is evident that those who are intent on harming the United States can do so quite effectively without relying on ballistic missiles. As many critics of NMD have noted, powerful bombs can be made within the U.S. from commercially available materials, and chemical and biological agents can be smuggled into the country in ordinary luggage. One can also imagine powerful computer viruses and other forms of unconventional attack.

It is evident that those who are intent on harming the United States can do so quite effectively without relying on ballistic missiles.

No doubt we will soon see an intense national debate on what measures will be needed to defend against future attacks of this sort. It is probably much too early to speculate on what might be required. But it is very doubtful that rapid construction of NMD will figure among the preferred remedies.

It would be premature at this point for anyone to come forward with a grand blueprint for America's future defense posture. Much time will be needed to analyze Tuesday's events and to identify the appropriate protective measures. But given the apparent irrelevancy of missile defenses to America's actual security situation, one of the first things that Mr. Bush should do as we begin redesigning our defense plans is to put NMD on indefinite hold.

8

A Missile Defense System Is Technologically Feasible

Ronald T. Kadish

Ronald T. Kadish, a lieutenant general in the U.S. Air Force, heads the Missile Defense Agency (MDA). Formerly called the Ballistic Missile Defense Organization (BMDO), the MDA is an agency within the U.S. Department of Defense that supports and conducts research and development of programs to defend the United States, deployed forces, and allies against missile attacks.

The United States faces a far different missile threat than it did in the Cold War, one that calls for missile defense. Intercepting and destroying a missile in space is a difficult but not impossible task. Researchers have developed kill vehicles with their own guidance systems that can hit and obliterate incoming missiles—a method called hit-to-kill. The American military has successfully tested several missile defense vehicles and has made great progress in overcoming the technical hurdles involved in developing effective missile defenses.

L et's consider the current debate about missile defense. Recalling what former Senator Sam Nunn [GA-D] said more than a decade ago, "We ought to do something unusual in Washington. We ought to let the facts speak for themselves."

Just what are the key facts?

The current missile threat

Fact One: The missile threat our Nation faces today is far different from the one we faced two or three decades ago. Yes, Russia still has the capability to launch a massive attack on the United States. But the likelihood of that happening has receded dramatically. Similarly, China has a limited but escalating ICBM [intercontinental ballistic missile] inventory. What is particularly worrisome, however, is the worldwide proliferation of ballistic missiles of all ranges, and of programs to develop weapons of mass destruction.

When the ABM [Anti-Ballistic Missile] Treaty was signed in 1972, there were only nine nations that had a ballistic missile capability. Today, almost three decades later, over 30 nations have such capability. Unfortunately, a number of these may pose a threat to the United States, to our allies, or to our troops overseas.

In spite of our counter-proliferation policies, our efforts have merely slowed, not stopped, this proliferation. Some nations have played a catalytic role in pushing these technologies, but the fact of the matter is that over the past 50 years the world's knowledge of missiles and weapons has blossomed.

What used to be on the shelves of a few select scientific libraries is now on the Internet, available to just about all who want it. What remains are the actual engineering challenges, the business of manufacturing and production, yet slowly but surely these are being overcome.

From rhetoric to reality

Fact Two: Intercepting a ballistic missile in space is a tough technical and management challenge—tough science and tough engineering—and has been ever since ballistic missiles were invented. But it is not impossible. We are now on the threshold of acquiring and deploying missile defenses, not just conducting research. We are, in fact, crossing over from rhetoric to reality, from scientific theory to engineering fact to deployed systems.

The first systems we will deploy provide close-in defense to intercept a short-range missile in its terminal phase. These include the Army's PAC-3, expected to be operational later this year [2001]. It is a greatly improved version of the Patriot missile that gained such notoriety in the Gulf. A few years later, we expect to deploy the Navy Area system that builds on existing fleet air defense capability. Both these systems are limited to defending a relatively small area.

We are now on the threshold of acquiring and deploying missile defenses, not just conducting research.

These systems will be followed by the Army's THAAD [Theater and High Altitude Area Defense System] and Navy Theater Wide systems later in the decade. These will both reach out farther back along the ballistic missile's incoming trajectory and so defend a wider region.

There's one inflexible rule about missile defense—the later you detect and intercept an enemy missile, the closer it will be when you destroy it, and the smaller the area you can defend. Conversely, the earlier you can detect, decide, and act, the farther away it will be when you destroy it, and the greater the area you can defend. In this business, farther is better; it gives you enough time to gain a chance for a second or third shot if you miss.

The national missile defense [NMD] system we've been working on destroys missile warheads in midcourse, the longest time of flight for a trajectory. Because it can reach so far out, it has the potential of defending a much larger area—in this case, the United States. To do this we employ space-based and ground-based sensors (the eyes, if you will, to see the launch and track the flight). We have a battle management system that interprets the information from the sensors, verifies if the missile is hostile, and determines the best point for intercept. Then, under human direction, it launches the interceptor, which we call a kinetic energy kill vehicle.

Hit-to-kill

Currently we have a small, 120-pound kill vehicle with its own guidance and sensing system. Once lifted into space and pointed in the right direction, on its final leg, it can steer itself into the target and obliterate the warhead by violent impact. That collision occurs at speeds of about 15,000 miles an hour or more, so there's not much left of the target reentry vehicle at those speeds.

You need patience to bring about revolutionary technology.

This is called hit-to-kill—hitting a bullet with a bullet. Traditional explosives don't work well in space, and the nuclear-tipped interceptors on which we relied 25 years ago—and on which the Russians still rely—have major political and operational drawbacks.

We've achieved success with hit-to-kill not once, but seven times out of the last 10 attempts in the past two years, with three different systems. Nevertheless, the development road for missile defense has been bumpy. Despite the successes, we still haven't achieved the degree of reliability we need. Each of the major programs is behind schedule. Our NMD program, for example, has had one hit followed by two well-publicized failures in our three intercept flight tests so far.

Patience is essential

Fact Three: You need patience to bring about revolutionary technology. Testing, by its nature, lends itself both to failures and to the successes that rise from their ashes. Wernher von Braun, who served as the project leader for the German V-2 program and later pioneered America's space and missile program, experienced many failures. He named these setbacks "successful failures" because he and his teams learned so much from them—a feature about high-tech development our critics tend to forget.

Do they remember that the Atlas ICBM program experienced 12 failures in its 2½-year flight-testing history; that the Minuteman-1 ICBM program suffered 10 failures in a 3½-year testing program; and that the Corona program, for our country's first operational reconnaissance satellite, survived 13 failures and mishaps before Discoverer 14 was orbited and its film recovered?

Yet through it all, patience by our leadership was essential—then and now—despite frustrations resulting from these technical difficulties. . . . to success.

Our test program follows a deliberate step-by-step approach that builds on simulations, ground testing, and risk reduction flights that don't involve an intercept attempt.

Missile defense is rocket science. We are on the threshold of solving the tough issues of defending against those ballistic missiles. We still have some stiff challenges ahead of us, but we are making remarkable progress. And while we have exceptionally talented and dedicated people working the demanding technical issues, we can't be successful without your patience and support.

9

A Foolproof Space-Based Missile Defense System Is Not Technologically Feasible

Theodore A. Postol

Theodore A. Postol, a longtime critic of missile defense, is professor of science, technology, and national security policy at the Massachusetts Institute of Technology (MIT).

Tests carried out on elements of America's missile defense system under development have revealed basic flaws that have gone unreported. The ability of rocket-powered interceptor missiles to locate and "kill" incoming enemy missiles can be easily compromised by the use of decoys and other countermeasures. Defense officials have carefully rigged tests by using only decoys that could be easily identified, and have possibly misled the public into thinking that significant progress is being made. The United States is wasting money on a space-based missile defense system that will not work.

O n June 23, 1997, a prototype of a U.S. military "kill vehicle" designed to intercept nuclear missiles lifted off from a launch pad on the South Pacific atoll of Kwajalein. Its purpose was not to seek out and destroy. Instead, it was to fly by and observe a group of objects that had been launched into space more than 20 minutes earlier from Vandenberg Air Force Base near Santa Barbara, California, almost 8,000 kilometers away—and determine whether it was possible to distinguish a cloud of decoys from the mock warhead they protected.

It was a big day for nuclear missile defense. Since the decoys used in this experiment were of very simple design, if the experiment showed that the warhead could not be reliably identified, it could mean the whole Star Wars defense plan [label given to President Ronald Reagan's 1983 call for developing a missile defense system] would for all practical purposes be unworkable, since the most primitive of adversaries could defeat it with the simplest of decoys. Of even greater importance, it would also be a clear demonstration of the fundamental physical reasons why any missile defense that relied on kill vehicles of this type could never be successful.

It worked—at least that's what we were told. But shortly after the experiment flew, three courageous people—a former employee of defense contractor TRW turned whistle-blower, a TRW retiree and a U.S. Department of Defense investigator—brought new evidence to light. Their information, coupled with my own investigation and repeated calls for a full accounting from U.S. representatives Howard Berman and Edward Markey, pointed to a different story—one of failure, a finding seemingly confirmed in February 2002 by a draft of a Government Accounting Office follow-on study, as reported by the journal *Science*. I believe that the top management of the Pentagon's Missile Defense Agency (previously known as the Ballistic Missile Defense Organization) and its contractors have misrepresented or distorted the results derived from the experiment and rigged the follow-on test program that continues to this day. These deliberate actions have hidden the system's critical vulnerabilities from the White House, Congress and the American citizens whom the missile defense program was supposed to protect.

How the defense system is supposed to work

As envisioned since 1996, the U.S. National Missile Defense effort consists of three main elements: infrared early-warning satellites, ground-based radars to precisely track warheads and decoys from thousands of kilometers away, and multistage, rocket-powered homing interceptor missiles launched from underground silos. The most critical element of this defense is the roughly 1.5-meter-long "exoatmospheric kill vehicle" that the homing interceptor deploys after being launched to high speed by its rocket stages. After deployment, the kill vehicle has about a minute to identify the warheads in a cloud of decoys as it closes on the targets at high speeds. To that end, it carries its own infrared telescope and has small rocket motors that enable it to home in on its prey. The kill vehicle does not carry a warhead. Rather, it is designed to destroy its quarry by force of impact.

The kill vehicle must be able to identify warheads and decoys without help from satellites, ground radars or other sensors.

When an enemy missile is launched, it typically takes 30 to 60 seconds to reach altitudes where the infrared early-warning satellites can detect the hot exhaust from its engines. These satellites orbit at an altitude of 40,000 kilometers and can be kept over the same point on the earth's surface. Once two or more detect the rocket, they can crudely track it in three dimensions by stereo-viewing. However, the satellites can only see the hot exhaust from the rocket's engines, so their tracking ends abruptly when the engines shut down—an event that typically happens in space at between 200 and 300 kilometers in altitude.

Roughly three minutes after engine shutdown, the rocket's upper stage and the just released warhead and decoys rise above the horizon, where they can be tracked by radar. The radar systems originally planned for this task operate on a very short wavelength (three centimeters at a frequency of 10 gigahertz), which allows them to identify objects to an accuracy of 10 to 15 centimeters from many thousands of kilometers away. This makes it possible to observe distinct reflections from different surfaces—even the seams on an object as it tumbles through space. The spacing and intensity of these signals, and the way their echoes vary as the orientation of a target object changes, can in some circumstances be used to determine which object is a warhead and which a decoy. If all goes well, this information will be used to deploy one or more interceptors within about 10 minutes of an attack's being launched. The interceptors will fly to the defense, destroying their targets about 18 minutes after launch.

That, at any rate, is how the system was initially supposed to work. President George W. Bush's latest proposal does not include this high-resolution radar, making tracking and identification of enemy missiles harder and delaying the interception time. But even with the more advanced original system, big problems surround the scenario. For starters, an adversary could alter the reflections from decoys and warheads by covering surfaces and seams with wires, metal foil or radar-absorbing materials. These simple strategies would render the radar unable to reliably sort out warheads from their armadas of decoys.

Compounding this problem is a simple fact: in the near vacuum of space, a feather and a rock move at the same speed, since there is no air drag to cause the lighter object to slow up relative to its heavier companion. This basic vulnerability makes it even easier for an adversary to devise decoys that will look like warheads to radar or an infrared telescope observing them from long range.

What's more, an adversary would likely deploy decoys and warheads close together and in multiple clusters. Under these conditions, even if the radar could initially identify a warhead among all the decoys, it couldn't track it accurately enough to predict the relative locations of the different objects when the kill vehicle encountered them some eight minutes later. Consequently, the kill vehicle must be able to identify warheads and decoys without help from satellites, ground radars or other sensors. If it cannot perform this task, the defense cannot work. This is where the infrared telescope comes in—and it was really this critical part of the system that the June 1997 test was all about.

How the kill vehicle identifies warheads

During a typical intercept attempt, the closing speed between the kill vehicle and targets is around 10 kilometers per second. If targets can be detected from a distance of 600 kilometers, that doesn't leave much time—a minute or less—to distinguish between warheads and decoys and maneuver to ram into the right target. The resolving power of the kill vehicle's telescope is quite limited, so all objects look like points of light. Still, the distinction can be made—by measuring the brightness of each object, and to some extent its wavelength or "color," which in turn can give clues to its infrared temperature.

If, for instance, one object is a tumbling, featureless sphere, no orientation will look different from any other, and its signal will be steady. However, if another object is of a different shape, the different faces it presents to the kill vehicle will show varying degrees of brightness as it tumbles end over end through space; a rod, for example, will be brighter when its more luminous side area is exposed to the telescope than when viewed end-on and will appear to the kill vehicle as a distant point of light that increases and decreases in brightness twice during each complete rotation. So if there is prior knowledge that one target is a tumbling rod and the other is a featureless sphere, it will be clear which is which.

The limited strategies available to the defense for distinguishing warheads from decoys put it at a disadvantage.

That's the theory. The truth is more complicated. For one thing, measuring temperature with this infrared equipment is not possible when objects in space are observed close to the earth, because their signals are routinely contaminated by reflected infrared radiation from the planet's surface; they are further confused by such factors as the amount of cloud cover, time of year and which part of the earth the target is over.

Even without such earthly interference, the limited strategies available to the defense for distinguishing warheads from decoys put it at a disadvantage. For example, one simple way for an adversary to make discrimination impossible is to put the warhead inside a balloon and deploy it with many additional balloons of different sizes and surface coatings. The temperature of a balloon exposed to the sun can be drastically altered, as can the amount of infrared heat it radiates and reflects from the earth and sun, depending on its size and surface coating. Balloons of different dimensions and with different coatings would each look slightly different. Since there would be no way to know why this was so, there would be no way to know which balloons were empty and which contained warheads—and discrimination by the kill vehicle's infrared telescope would be impossible.

This is the central point that backers of missile defense have not been able to circumvent.

Problems with the tests

So far, there have been seven tests, the most recent last December [2001]. In each case, a payload of targets has been launched by a modified Minuteman II intercontinental ballistic missile (ICBM) from Vandenberg Air Force Base toward Kwajalein. The target-carrying Minuteman completes powered flight in about three minutes and deploys a rocket-powered vehicle called the Multi-Service Launch System. This vehicle takes another four and a half minutes to deploy its payload, but only after rotating nearly 90 degrees so it can release the targets along a single downward direction in space. Since the kill vehicle telescope has a field of view roughly equal to that of a person looking through a soda straw (about one to two degrees), the payload deployment along a single direction assures that the targets will all be in its limited sights when they arrive at Kwajalein some 20 minutes later. Since simultaneous observation of targets is critical to quickly distinguishing decoys from warheads, this specialized deployment geometry gives the kill vehicle a significant advantage—one it is hardly likely to have in a real-world attack. (If, instead, the targets were deployed in many directions, the kill vehicle would have to slew between many clusters of targets, viewing each for tens of seconds to get the same readings. Even if it could identify the right target, there would likely not be time to maneuver and intercept.)

When the first flight test was performed, 10 targets were to be observed by the kill vehicle. These included a roughly two-meter-long, spin-stabilized mock warhead; two cone-shaped rigid decoys that were of roughly the same shape and size as the mock warhead: four spherical balloons (two with a diameter roughly equal to that of the base diameter of the mock warhead, and two about half that size); a small cone-shaped balloon; a large spherical balloon; and the upper rocket stage that deployed the decoys and warhead.

In spite of the numerous and fundamental experimental failures in the first trial, . . . the Defense Department reported that the experiment was an unqualified success.

At first glance, it might seem that this ragtag collection of decoys is just what an enemy would throw at us. But since the makeup of these objects and the space-infrared environment in which they operated were fully known—all the tests have been carried out around the same time in the early evening, assuring that the geometry of the sun and earth are essentially the same in every experiment—it was possible, at least in principle, to predict how each would look to the kill vehicle. The predictions indicated, for instance, that the two medium balloons would have scintillating signals as bright as that of the spin-stabilized mock warhead, which had roughly the same diameter. Each of the rigid cone-shaped decoys was expected to look like a tumbling warhead. The large balloon and upper rocket stage were expected to look much brighter than all the other objects, while the small spherical balloons and the cone-shaped balloon would stand out for their dimness. Under these simplified conditions, and with detailed prior knowledge of the characteristics of each object, it must have seemed quite likely that the kill vehicle could pick out the "warhead" from among the decoy companions.

The results of the actual trial were quite unexpected, however, and must have been extremely disconcerting to then director of the Ballistic Missile Defense Office Lt. Gen. Lester L. Lyles and his engineering team. Lyles reported that the trial had proven that discrimination of warheads in a cloud of decoys was possible. However, we now know there was a serious basic problem with the first integrated flight test that would likely, even with the targets' expected characteristics known prior to launch, make any of the data gathered by the kill vehicle essentially useless.

To begin with, one of the medium-sized balloons failed to fully inflate, resulting in its looking half as bright as expected. The fluctuation characteristics of the mock warhead's signal, meanwhile, changed over time, making the probability of its being the warhead appear at different moments more than five times higher or lower than expected. Indeed, the fluctuation characteristics of all the objects were either substantially different from the predictions or changed in time so drastically that if they could be matched to the template of expected values at one time, they could not be matched to the same template even seconds later.

It makes no sense to advocate a concept that will not work.

That was bad enough. But the real problem was that the kill vehicle's main infrared sensor failed to cool to its 12-degree-Kelvin design temperature, achieving instead a temperature no lower than 13.5 Kelvin. This difference is the same as if a space suit had been designed to keep an astronaut in a temperature environment of 20 °C but instead put her in an environment of 66°C.

Since the sensor was very hot relative to its design operating temperature, the measured target signals were contaminated with heat-generated electronic noise. The unexpectedly high temperature also caused unpredictable changes in the efficiency with which each of the tens of thousands of tiny, independent infrared sensors in the kill vehicle converted infrared signals to electronic. These sensors are arrayed at the telescope's "eyepiece," so that an electronic image of the instrument's field of view can be formed, much the way images are formed by solid-state TV cameras.

Since each infrared sensor's performance was different in detail from the performance of the other sensors, and since the details of how the performance of each sensor changed with the unexpectedly high temperature were unknown, it was not possible to accurately measure the brightness of distant targets, or even the brightness of these targets relative to each other. Because knowing the brightness of a target is critical to identifying it, the singular fact that the main sensor was not at its operating temperature, and that the performance of each of its tens of thousands of elements was unknown, means that the kill vehicle's capacity to discriminate its target was severely compromised.

To understand this basic point, imagine that an object is being observed by two infrared sensors with different conversion efficiencies. When light from the object strikes one of the sensors, a certain brightness is recorded. When it strikes the other sensor, a different brightness is recorded. Unless the conversion efficiencies of both sensors are known, the actual brightness of the object cannot be determined.

This well-known problem and others associated with it would typically have been dealt with by calibrating the performance of each sensor over the range of expected operating temperatures prior to the experiment. However, the experimental team had not performed calibration measurements on the array at 13.5 Kelvin and higher because it had not expected such a massive failure in the sensor's cooling system. Additional sensor calibration data was supposed to have been obtained by observing an infrared star of known brightness, Alpha Bootes, but the noise in the many sensor elements, and changing sensor array temperature during the test, rendered this measurement useless.

So, I believe, when the carefully contrived test failed, the true results of the experiment were hidden through careful selection of the data used in the analysis—and the way in which those data were interpreted. The kill vehicle collected about 63 seconds of data, starting at a range of roughly 460 kilometers and continuing until it flew by the targets at a speed of 7.3 kilometers per second. The first 30 seconds of data were so severely contaminated by heat-generated electronic noise that none of them could be used in the postflight analysis. For various other reasons— some scientifically legitimate, but also including the fact that one of the medium-sized balloon decoys suddenly began to look more and more like the warhead—the last 16 seconds of the flyby were also removed.

That leaves the data collected during the 17-second period between 33 and 16 seconds prior to flyby as the only data officially reported by the contractor. The first five seconds of this period were eventually excluded as well, because changes in both the measured brightness and the fluctuation in brightness of each target caused three different targets to look like the warhead during this short interval. The remaining 12 seconds, then, were the only time when the signals were sufficiently stable that the observed data could be matched to a template of expected warhead and decoy characteristics. But because the sensor measurements involved unknown conversion efficiencies, and it was therefore impossible to use the original template, a new template was created after the test to fit the uncalibrated sensor data. It was this after-the-fact template, matched to almost certainly inaccurate measurements, that formed the basis of the claims about the experiment's success.

Such claims, it almost goes without saying, are meaningless.

Insisting on success

As I have noted, in spite of the numerous and fundamental experimental failures in the first trial, TRW and the Defense Department reported that the experiment was an unqualified success.

A second, similar test was launched on January 16, 1998—and once again, fundamental signs of the system's inadequacy continued to be overlooked. Chief system architect Keith Englander claimed that in both tests "we were able to pick the reentry vehicle out of the target complex." Lieutenant General Lyles and his successor, Lieutenant Gen. Ronald Kadish, also praised the experimental results before Congress. Kadish went so far as to assert that the first two experiments had "demonstrated a robustness in discrimination capability that went beyond the baseline threat." The Lincoln Laboratory scientists who helped review the experimental claims for the Department of Defense after Nira Schwartz, the TRW whistle-blower, had raised the alert made no mention of the sensor array problems in their public report, issued in late 1998.

Between mid-1998 and December of 2001, five other trials were conducted. The decoys that were the most difficult to discriminate from warheads in the first two tests were removed from these and all subsequent missile defense development tests. These included the cone-shaped decoys that had the same size and appearance as the mock warhead, the striped balloons with the same base diameter as the warhead and the small cone-shaped balloons that could easily be made to look like warheads if their surface coating and/or dimensions were slightly altered.

The only "decoy" flown in the three tests immediately following the first two trials was a very large balloon, which was easily identifiable because it was known prior to the test to be seven to 10 times brighter than the mock warhead. When the seventh test was ultimately flown, last December 3, the diameter of the large balloon was reduced somewhat—from 2.2 meters to 1.6 meters—but it was still three to five times brighter than the warhead. And for future trials, according to accounts in the *New York Times,* a completely new set of infrared decoys is to be unveiled. These are to be made up only of spherical balloons composed of uniformly unvaried materials and without stripes, virtually guaranteeing that they will have perfectly steady and unvarying signals. By contrast, the dummy warheads will intentionally be deployed so as to tumble end over end. This simulates the most primitive ICBM technology, where the warhead is not spin-stabilized—so as to maintain its orientation in space and make its entry into the atmosphere and subsequent flight path more predictable—and causes its signal brightness to scintillate wildly.

The implication of this carefully contrived choice of new decoys is chillingly clear. All the problematic shortfalls in the defense system discovered in the first two experiments have been removed through the painstaking designing of a set of decoys that would never be used by any adversary, but *would* make it possible to distinguish warheads from decoys in flight tests.

This should be of profound concern to every U.S. citizen. The officers and program managers involved in developing the antimissile system have taken oaths to defend the nation. Yet they have concealed from the American people and Congress the fact that a weapon system paid for by hard-earned tax dollars to defend our country cannot work.

How a successful missile defense system might work

Whether or not one believes there is any threat serious enough to require deployment of a national missile defense, it makes no sense to advocate a concept that will not work. There is a way, though, to provide a defense that would likely be highly effective, a strategy that avoids the serious and as yet unsolvable problems posed by space-deployed decoys that I have discussed.

A "boost-phase" missile defense would target intercontinental ballistic missiles in their first few minutes of flight, while they are still being accelerated up to speed by their rocket engines. Because such a system would consist of very fast, short-range (perhaps a thousand kilometers) interceptors positioned only a few hundred kilometers from the "rogue" nations likely to attack the United States, it would be effective only over a relatively small region of the earth. While the system would be devastating when used against geographically small emerging missile states, it would be largely useless against missiles launched from vast countries such as Russia or China; it would simply not be feasible to position enough interceptors close enough to their launch sites. This is good news too, however, since it would allow the U.S. to target the Third-World states it claims to be most concerned about without provoking negative reactions from Russia and China.

In the case of North Korea, ships or converted Trident submarines could serve as launch platforms for these interceptors. Silos deployed in eastern Turkey would be effective for covering launches from inside Iraq. If a defense were required against Iran, its larger size and location would require defense sites in Turkey, Azerbaijan, Turkmenistan or the Caspian Sea.

The current U.S. approach to missile defense is a direct outgrowth of the irrational idea that "we" can deal with the world without working with others.

When an ICBM was launched, it would be detected and tracked by sensors on the ground, in unmanned aircraft, aboard ships or on satellites. The interceptors would accelerate to 8 to 8.5 kilometers per second in a little over a minute. At these speeds, even if their launch were delayed for a minute or more in order to establish the enemy missile's trajectory, the interceptors could still destroy the ICBM while it was in powered flight, causing its warhead to fall far short of its target.

Unlike the proposed space-based system, this defense would be difficult to counter. Countries seeking to defeat it might try to reduce the boost-phase flight time, thereby narrowing the window of opportunity for a successful intercept. But that would require the development of highly advanced solid-propellant ballistic-missile technology—an innovation that is in a completely different league than the liquid-fuel, Scud missile technology that is currently the foundation for the missile programs of North Korea, Iran and Iraq. In addition, the technology needed to implement this defense is far less demanding than that needed for midflight intercepts in space. Because boost-phase interceptors would only need to detect the very hot plume of the booster and not the cooler warhead or decoys, such interceptors could use higher-resolution short-wavelength sensors that are easier to build and much less costly than the long-wavelength sensors used by the exoatmospheric kill vehicles of the planned nuclear-missile defense system. Finally, the ICBM booster target is large and would be destroyed by a hit almost anywhere, so the probability of a successful intercept would be very high.

Some boost-phase defense systems would certainly face significant geopolitical obstacles. Getting countries such as Azerbaijan or Turkey, for instance, to allow basing of interceptors in their territory could be a challenge. If a deployment against Iran were needed, it would also require close cooperation between Russia and the United States, which would likely increase existing Chinese concerns about a U.S.-Russia alliance.

However, these and other problems are all far more manageable than those raised by the currently planned space-based nuclear-missile defense system. Even the first phase of this fragile and easily defeated defense is threatening to create serious problems with both Russia and China—

while providing the U.S. with essentially no meaningful protection against them or any other potential enemy state.

A plea for scientific and political leadership

In the wake of the terrifying attacks on the World Trade Center and Pentagon, the entire civilized world will need to work to defeat the forces of ignorance, intolerance and destruction. In my view, the current attitude of the Bush administration that "we can go it alone" is one of the most dangerous and ill-considered security policies to be adopted and pursued by the United States in recent memory.

The current U.S. approach to missile defense is a direct outgrowth of the irrational idea that "we" can deal with the world without working with others. It is not only an irrational position when examined in terms of social realities, it is also irrational in terms of basic principles of physical science. It is sad and disturbing that the most technologically advanced and wealthy society in human history has displayed so little scientific and political leadership on matters that will almost certainly affect every aspect of global development in the 21st century.

10

America Was Right to Abandon the Anti-Ballistic Missile Treaty

Jesse Helms

Jesse Helms is a Republican senator from North Carolina and former chair of the Senate Foreign Relations Committee.

President George W. Bush made the right decision when he announced in December 2001 that the United States would be withdrawing from the Anti-Ballistic Missile (ABM) Treaty. The changing world situation, including the fall of the Soviet Union and the growing threats posed by Iran, North Korea, and other rogue nations, made the treaty obsolete. The treaty would have severely constrained America's ability to test and develop an effective missile defense system.

Opponents of President George W. Bush's decision to withdraw from the 1972 Antiballistic Missile Treaty have called it the cornerstone of strategic stability. Yet the foundation beneath that cornerstone collapsed 10 years ago along with the Soviet Union. Today, with the proliferation of nuclear weapons and missile technology, along with multiple threats to our homeland, blind adherence to the ABM treaty could make it our national gravestone.

Rogue nations threaten America

Since the collapse of the Soviet Union, rogue nations such as Iran, Iraq and North Korea, ravenous for any advantage against the U.S., have feasted on opportunities to develop nuclear weapons and ballistic missiles. These regimes already may be able to launch a missile that could strike the U.S. or our allies with a nuclear warhead. They will certainly have that capability in a few years.

As long as our nation is vulnerable to nuclear threats from the likes of [North Korean leader] Kim Jong Il, [Iraqi leader] Saddam Hussein and the ayatollahs of Tehran, we also will be vulnerable to blackmail, intimidation and other efforts to influence our foreign policy.

The bottom line is that the ABM treaty is prohibiting important tests of our most promising [missile defense] systems.

The ABM treaty was created when the U.S. had one real enemy: the Soviet Union. But deterrence alone is no longer an effective strategy when the world is tainted by unpredictable tyrants and terrorists who are difficult to target and have little concern for human life, or when an unauthorized or accidental launch of a missile occurs. Yet some Chicken Littles in Washington warn that the sky will fall now that the president has decided to withdraw from the treaty, saying that China will be forced to build more nuclear missiles and an arms race will ensue in South Asia. This argument doesn't pass the laugh test. Communist China has been modernizing its nuclear forces for years and will continue to do so regardless of our missile defense policy. Worse, it is Beijing that is spreading instability by proliferating dangerous technologies to Pakistan—in violation of international agreements—and to rogue states in the Middle East.

A waste of money?

Others argue that missile defense is a waste of money, that the technology won't work and that we should spend scarce dollars on other threats.

This year [2001], the federal government spent less than $2 billion on national missile defenses, which is one-fifth of what we spend annually on counter-terrorism and less than 1% of the Pentagon's annual budget. The president's request added about $3 billion to these efforts. A series of tests—including one Dec. 3 [2001]—has proven our ability to "hit a bullet with a bullet."

Some have urged the president not to scrap the treaty but to "stretch" it so that testing and development of missile defenses can go forward while the treaty remains intact. This is not plausible. The treaty severely constrains testing and contains no provisions to allow violations of the treaty by mutual consent. The bottom line is that the ABM treaty is prohibiting important tests of our most promising systems and impairing our ability to have a rudimentary missile defense ready by 2004—when the threats posed by some rogue states are expected to mature.

Our relationship with Russia

President Bush has done an extraordinary job of convincing the Russians of the shared threats we face. Several senior Russian officials—including President Vladimir V. Putin—have already stated the U.S. decision doesn't threaten Russia's security, won't affect our mutual efforts to make deep cuts in their nuclear stockpile and will not undermine Russia's new relationship with the U.S. Most of our allies recognize these facts and do—or will—support this decision.

The [September 11, 2001, terrorist attacks on America] demonstrated that we shouldn't underestimate our enemies and that we must defend against all threats. The president has done the right thing by announcing our withdrawal from the treaty and setting U.S.-Russia relations on a new and positive strategic footing.

Congress should not pursue political chicanery to obstruct him.

11

America Should Not Withdraw from the Anti-Ballistic Missile Treaty

Salih Booker

Salih Booker is the executive director of Africa Action, an advocacy group on African affairs.

President George W. Bush made the wrong decision when he announced in December 2001 that the United States would be withdrawing from the Anti-Ballistic Missile (ABM) Treaty. The decision reflected the influence of unilateralists in the Bush administration who oppose virtually all international treaties or systems of collective security. By discarding the ABM Treaty, America signals its contempt for world opinion and makes it impossible to engage other nations in diplomatic arms control efforts.

George W. Bush's abrogation of the Anti-Ballistic Missile (ABM) treaty on December 13 [2001] was a reckless act taken without regard for the consequences—it has dealt a severe blow to the idea of a world order grounded in collective security. Bush justified this act of hubristic contempt for the rest of the world as a measure to protect the American "homeland," but it actually will increase the danger of nuclear war and place this country at greater risk.

The abandonment of the ABM treaty represents a victory for Donald Rumsfeld's Defense Department over Colin Powell's State Department. It's significant that the point man in the ABM negotiations was John Bolton, under secretary for arms control and international security, a Rumsfeld loyalist planted in the State Department over Powell's objections. At Bolton's confirmation hearing, Jesse Helms told him, "John, I want you to take that ABM treaty and dump it in the same place we dumped our ABM treaty co-signer, the Soviet Union, and that is to say, on the ash heap of history." Mission accomplished. Bolton divides US policy-makers into two opposing camps: the "Americanists" and the "globalists," the latter of which would entangle us in treaties. With the ABM pullout, the Bolton-

From "Mission Unilateralism," by Salih Booker, *Nation*, January 7, 2002. Copyright © 2002 by The Nation Company, L.P. Reprinted with permission.

engineered refusal to sign the biological warfare treaty, the opposition to the International Criminal Court and US subordination of multilateral relief and reconstruction projects in Afghanistan to pursuit of the war, the "Americanists" are now in the saddle—and with them the Rumsfeld/Paul Wolfowitz doctrine of a wider war on terrorism, which could embroil the United States in a cycle of bloody overseas interventions. Rumsfeld told NATO it should "prepare now for the next war."

Global reaction

Reacting to Bush's withdrawal, [Russian president] Vladimir Putin called it a "mistake" but indicated that the US-Russian relationship was strong enough to survive this setback. Yet Putin is surrounded by political elites who are deeply distrustful of Washington. They are already reminding Putin of the despised "Gorbachev-Yeltsin syndrome"—a pattern of far-reaching Russian concessions in the 1980s and 1990s that were met by broken Western promises. As Aleksei Arbatov, deputy chair of Parliament's defense committee and a leading pro-Western politician, put it, "After the tragedy of September 11, Russia extended its hand full length to meet the US in the spirit of cooperation and even mutual alliance. . . . Today, the US has spat into that extended hand."

Bush has jettisoned it [the ABM treaty] without anything to replace it.

Bush's cavalier dumping of the treaty may make it impossible to engage Russia, China, India and Pakistan in a sustained diplomatic effort to outlaw nuclear, biological and chemical weapons. Now China may decide to upgrade and expand its nuclear force. This prospect seems not to have troubled the Bush team; nor was there any apparent concern about the impact of a Chinese buildup on India's and Pakistan's ominous nuclear calculations, especially with tensions between India and Pakistan at a rolling boil because of the [December 2001] terrorist attack on the Indian Parliament.

The Administration may be right in saying that the ABM treaty is something of an anachronism, a product of an outmoded bipolar cold war system of nuclear deterrence. But the treaty symbolized US willingness to be bound by international agreements aimed at closing off one of the sources of a nuclear arms race. It represented an understanding that America's security comes only from common security. Now Bush has jettisoned it without anything to replace it. For the hugely expensive, highly uncertain antimissile system the Administration hopes to build does not deal with two primary nuclear perils haunting the world. One is the degraded state of Russia's nuclear weapons system, its platoons of unpaid nuclear scientists, its poorly guarded uranium stockpiles and the growing possibility of an accidental launch. The other danger comes not from "rogue nations" but from freelance terrorist groups like Al Qaeda acquiring nuclear weapons, which they could deliver in small planes, ships or suitcases.

An imperial presidency

Unilateralism abroad mirrors a new burst of unilateralism at home, exemplified by Bush's gathering to himself the powers of an imperial presidency, for example his presidential order setting up military courts to try terrorists. The ABM pullout was another expression of this aggrandizement. Constitutional scholars like Bruce Ackerman of Yale Law School question whether a President has the right to cancel a treaty without Congressional approval—after all, he needs two-thirds of the Senate to approve one. Throughout US history two Presidents sought Congressional advice and consent before reneging on a treaty, and a third, Jimmy Carter, unilaterally withdrew from one, but his action triggered a constitutional challenge by conservative senators. The Supreme Court refused to hear that case, ruling that it was a political question. That leaves it to Congress to stop future withdrawals. What if a President decided to resign from the United Nations? Congressional Democrats ought to be raising hell about Bush's constitutional bypass.

12

A Missile Defense System Would Not Enhance America's National Security

Steven Weinberg

Steven Weinberg, a Nobel Prize winner in physics, has been a consultant to government agencies on national defense issues. His writings include The First Three Minutes *and* Dreams of a Final Theory.

America does not face a fixed missile threat. Attempting to construct a national missile defense system that would take years to fully deploy may provoke other nations into increasing their nuclear arsenals, making America less secure. Pursuing missile defense may also weaken America's national security by taking away funds from other defense and security programs, such as programs to prevent nuclear proliferation. The employment of a missile defense system would also harm American relations with other countries.

Would a national missile defense system of the sort proposed by the Bush administration help or hurt our national security?

At first sight, this question seems to answer itself. Isn't any missile defense, however ineffective, better than no missile defense at all?

No fixed threat

One trouble with this reasoning is that we do not face a fixed threat, a threat independent of what we do about missile defense. True, we are not now in the position we were in during the 1960s and 1970s, when we could reasonably expect that any US missile defense system would be countered by an increase in Soviet offensive missile forces. The current Russian economy would not support an increase in Russia's missile forces and, indeed, the Russians have been eager to reduce their forces, reportedly down to some two thousand or so strategic nuclear warheads. But this is still a force that could destroy the US, and much else in the world besides. The large size of their arsenal also increases the danger that Russian nuclear weapons or even long-range missiles might be stolen or sold to terrorists or rogue states. I am told that Russia now maintains tight control over its strategic nuclear weapons, but this wasn't true in the early 1990s and it may not be true in future. There is nothing more important to American security than to get nuclear forces on both sides down at least to hundreds or even dozens rather than thousands of warheads, and especially to get rid of MIRVs [multiple independently targeted reentry vehicles], but this is not going to happen if the US is committed to a national missile defense.

The Russian nuclear force is the sole remnant of its status as a superpower. Whatever good feelings may exist now between us and Russia, any US system that might defend our country against even a few Russian intercontinental ballistic missiles therefore sets a limit below which the Russians will not go in reducing their strategic nuclear forces. Even if Russia is forced by economic pressures to continue reducing its missile forces, it can cheaply maintain its deterrent, although in ways that are dangerous. It could, for example, remove whatever inhibitions it may now have from launching its missiles on a moment's notice. Nor is Russia likely to eliminate its MIRVs if the US goes ahead with missile defense. . . .

As for China, it has right now about twenty nuclear-armed intercontinental ballistic missiles, enough for a significant deterrent against any attack from Russia or the US. A National Intelligence Estimate that was leaked to *The New York Times* and *The Washington Post* predicted that if the US develops a national missile defense, then the Chinese will increase their forces from about twenty to about two hundred missiles. And if China makes this sort of increase in its missile force, then what will Japan and India do? And then what will Pakistan do?

It may seem contradictory to argue that the proposed national missile defense system would probably be ineffective against even a small attack by a rogue state, while also arguing that it would prevent needed reductions in Russian missile forces and promote increases in Chinese missile forces. But each country "prudently" tends to overestimate the effectiveness of any other country's defenses, especially as they may develop in future. The Soviet deployment of a primitive antimissile defense of Moscow was a major factor in America's decision to multiply its warheads by deploying MIRVs, so that Moscow was in more danger after it was defended than before. I remember how in the early 1970s US defense planners became terrified that Soviet antiaircraft missiles might be given a role in defense against intercontinental ballistic missiles, something that never happened. Imagine then how Russians and the Chinese defense planners will take account of the unilateral American withdrawal from the 1972 ABM treaty.

The choice we face

If it were possible tomorrow to switch on a missile defense system that would make the US invulnerable to any missile attack, then I and most other opponents of missile defense would be all for it. But that is not the choice we face. What is at issue is a missile defense system that will take almost a decade to deploy in its initial phase, and then many more years to upgrade to the point where at best it would have some effectiveness against some plausible threats. During all of this time, however, American security will be damaged by measures taken by Russia or China to preserve or enlarge their strategic capability in response to our missile defense.

We do not face a fixed threat, a threat independent of what we do about missile defense.

There is one sort of missile defense that would not raise these problems. . . . A defense that targets intercontinental ballistic missiles during the boost phase with either missiles or airborne laser beams could only defend against missile launches within a limited geographical area; and it would also be immune to decoys and other penetration aids. We could defend against the launch of North Korean missiles by using short-range missiles based on ships in the Sea of Japan, though to defend against a launch from Iraq or Iran would require cooperation from Turkey or some republic of the former Soviet Union, respectively. Such a defense would have no effectiveness against missiles launched from sites in Russia or China, which during the boost phase would be beyond the range of any missiles or airborne lasers we might deploy. For this reason, although this sort of defense would violate the 1972 ABM treaty, [Russian] President [Vladimir] Putin has already indicated that he would consider revising the treaty to allow it. But a boost phase intercept system would have all the destabilizing effects of other missile defense systems if it were based on satellites, or if it were combined with exo-atmospheric midcourse interceptors like those of the Clinton-Bush National Missile Defense proposal.

Foreign relations

Developing a national missile defense system would also harm our foreign relations. It would add to the general perception that the US is unwilling to be bound by international agreements, such as comprehensive test ban treaties or environmental agreements. It would weaken Putin's hand in dealing with Russian ultranationalists. By trying to defend the US from missile attacks while leaving our allies defenseless, missile defense would tend to undermine alliances like NATO. A boost-phase intercept system would not really be an exception; it is true that the interception of a long-range missile in the boost phase does not depend much on the destination of the missile, but by interrupting the boost it would probably only cause the warhead to fall short, perhaps on an ally, such as Canada or Germany.

A missile defense system would hurt our security in another important way, by taking money away from other forms of defense. We are simply unable to do everything we can imagine that might defend us. We need to upgrade our hospitals to deal with biological attack; improve security along our border with Canada and in our ports; upgrade the FBI computer system; and so on. All of our activities along these lines are constrained by a lack of funds. Legislation to increase funding for homeland defense was blocked in the House of Representatives because the amounts of money requested exceeded the administration's guidelines.

If we are particularly worried (as we should be worried) about terrorist nuclear attacks on the US, then we ought to give a very high priority to working with Russia and other countries to get rid of the large stocks of weapons-grade plutonium and uranium that are produced by their power reactors. Russia now holds about 150 tons of plutonium and one thousand tons of highly enriched uranium. This material could be used not only to make nuclear bombs, which can be delivered to the US in all sorts of ways; even a technically unsophisticated terrorist could instead use it to make so-called "dirty" bombs, in which an ordinary high explosive is surrounded with highly radioactive material that when dispersed in an explosion would make large urban areas uninhabitable.

Developing a national missile defense system would . . . harm our foreign relations.

Unfortunately, this material is not under tight control. Since 1991 there has been a bipartisan Nunn-Lugar Cooperative Threat Reduction Program that among other things aims at improving the security of Russian control over fissionable materials and making Russian plutonium and uranium unusable as nuclear explosives, but this too is not being adequately funded. A bipartisan panel headed by Howard Baker and Lloyd Cutler has called for spending at an average level over the next decade of about $3 billion a year for securing, monitoring, and reducing Russian nuclear weapons, materials, and expertise. The amount in the 2002 budget for these activities is only about $750 million, even after substantial increases by Congress. Comparison of these figures with the $60 billion quoted cost (certain to be greatly exceeded) of a minimum missile defense system gives a powerful impression that the Bush administration and some in Congress are not entirely serious about national security.

American fascination with space

I was at a press conference in Washington in November 2001, when the Federation of American Scientists released a letter signed by fifty-one Nobel laureates that opposed spending on national missile defense programs that would violate the 1972 ABM treaty. One of the reporters present asked me why, if the arguments against national missile defense are so cogent, many people in and out of government are for it? It was a good question, and one to which I am not sure I know the answer. There are the usual pressures for large military programs that come from defense contractors and from politicians trading on patriotism. The arguments for national missile defense may seem simpler and more straightforward than the arguments against it. But I think there is also a peculiar fascination with anything that projects American power into space. How else explain the idiocy of the International Space Station, or the card tables that were set up at airports during the Reagan administration by people advocating a "high frontier" missile defense program? I have to admit that thoughtlessness is not a monopoly of missile defense advocates. Some opponents of missile defense are automatically against any large military program. In assessing missile defense, or anything else for that matter, there is no substitute for actually thinking through the issues.

In my own field of physics, we make a distinction between applied physics, which is motivated by some social need, and pure physics, the search for knowledge for its own sake. Both kinds of physics are valuable, but not everything pure is desirable. In seeking to deploy a national missile defense aimed at an implausible threat, a defense that would have dubious effectiveness against even that threat, and that on balance would harm our security more than it helps it, the Bush administration seems to be pursuing a pure rather than applied missile defense—a missile defense that is undertaken for its own sake, rather than for any application it may have in defending our country.

13

A Missile Defense System Would Enhance American Power

Lawrence F. Kaplan

Lawrence F. Kaplan is a senior editor at The New Republic, *a journal of opinion. His writings about foreign policy have appeared in* Commentary, *the* Washington Post, *and other publications.*

The United States needs missile defense not just for protection against attack, but to protect its ability to use force in international relations. Currently, countries that possess missiles have largely succeeded in placing themselves off-limits for forceful American intervention—a situation the United States should and could prevent by constructing a robust missile defense. A sea-based missile defense system is preferable to the Alaska-based one proposed by President Bill Clinton and continued by President George W. Bush.

Thank God for missile defense. For Washington foreign policy types who spent the last decade snoring through panels at the Brookings Institution, the salad days are here again. Just when it seemed no one cared about national security issues, back comes missile defense, and it's as if the Star Wars debates of the 1980s never ended. . . .

But what really makes it feel like the 1980s is that in an era when ideology has been banished from most foreign policy debates, ideologues have made this one their exclusive property. Overnight, all the Reagan-era battle lines have reappeared. On one side is the old Zabar's consensus, featuring the *New York Times* editorial page, *The New York Review of Books*, and a parade of New School professors. They've brought back all the Reagan-era arguments: Missile defense will destabilize the international scene and spur a new arms race, and it won't even work. On the other side, the loudest clamor for missile defense comes from a chorus of congressional yahoos who see in the program an opportunity to erect Fortress America. If we can build a shield to protect the United States from attack,

From "Offensive Line," by Lawrence F. Kaplan, *The New Republic*, March 12, 2001. Copyright © 2001 by The New Republic, Inc. Reprinted with permission.

the argument goes, we won't have to send troops abroad in search of dragons to slay. As it happens, all this ideological posturing bears little relation to the world in which we now live. In fact, the strategic logic of missile defense runs entirely counter to the claims of isolationist champions and liberal critics alike. The real rationale for missile defense is that without it an adversary armed with long-range missiles can, as Robert Joseph, President Bush's counterproliferation specialist at the National Security Council (NSC), argues, "hold American and allied cities hostage and thereby deter us from intervention." Or, as a recent RAND study on missile defense puts it, "[B]allistic missile defense is not simply a *shield* but an *enabler* of U.S. action." In other words, missile defense is about preserving America's ability to wield power abroad. It's not about defense. It's about offense. And that's exactly why we need it. . . .

History of the ABM debate

The debate over anti-ballistic missile (ABM) systems began not in the Reagan era but in the Roosevelt era, with the appearance of the first operational ballistic missile, the German V-2, late in World War II. (After being propelled outside Earth's atmosphere by rocket engines, ballistic missiles rely on gravity.) Within a year of the war's end, the Pentagon launched two programs to explore ways to counter the threat. By the mid-'50s, first the Air Force and then the Army had devised ABM proposals that would combine long-distance radar with nuclear-tipped interceptor rockets. The Army moved forward with its ABM program in the 1960s and by decade's end was set to begin construction. (A missile defense system was, in fact, briefly deployed—guarding a single missile site in North Dakota—before being scrapped in 1974.)

The strategic logic of missile defense runs entirely counter to the claims of isolationist champions and liberal critics alike.

During the same decade, however, Robert McNamara, defense secretary under [presidents] John F. Kennedy and Lyndon Johnson, undermined the case for missile defense by enshrining in official policy a version of deterrence theory, Mutually Assured Destruction (MAD), which held that the surest way to avert nuclear holocaust was for the Soviet Union and the United States to remain vulnerable to each other's arsenals. Consequently, MAD's defenders deemed anything that diminished this mutual vulnerability—particularly missile defense—a threat to stability. It was a curious argument, and not everyone bought it. Strategists like Albert Wohlstetter and Herman Kahn continued to make the case for missile defense, as did scientists like Edward Teller, who argued that it was better to "shoot at enemy missiles than to suffer attack and then have to shoot at people in return." Nonetheless, after years of bitter debate, much of it poisoned by the toxic residue of Vietnam, the McNamara logic prevailed. In 1972 the Nixon administration signed the Anti-Ballistic Missile Treaty with Moscow, effectively banning national missile defense (NMD).

The Star Wars debate of the 1980s, set off by President Reagan's proposal to build space-based defenses, basically amounted to a rehashing of "the great ABM debate" of the 1960s. But the technology had grown more sophisticated and the arguments more crude. Strategists and scientists faded into the background, supplanted by Republican revolutionaries and bien-pensant leftists. Nothing was ever deployed, which was just as well, since the technology wasn't there and, even if it had been, the Soviets could have easily overwhelmed it with their huge arsenal. And then the USSR crumbled, ending the argument. Now, with President [George W.] Bush's pledge to deploy an ABM system, the debate enters its latest installment. Only this time something's different: the world.

A changed world

Aside from the absence of Soviet communism, the main thing that's different is that more countries possess ballistic missile technology. In the past few years alone, India and Pakistan have set off a combined total of twelve atomic explosions; Pakistan, Iran, North Korea, and China have test-launched ballistic missiles; Iraq, Syria, and Libya have reportedly acquired missile components; and both China and Russia have continued to export ballistic missile technology throughout the Middle East. Still, U.S. policymakers have been slow to recognize the danger. As late as 1998, Joint Chiefs Chairman Henry Shelton averred that "the intelligence community can provide the necessary warning" if one of these countries was developing "an ICBM threat to the United States." Alas, just a week after Shelton's pronouncement—and with no warning whatsoever from the intelligence community—North Korea demonstrated its intercontinental ballistic missile (ICBM) capability by launching a three-stage rocket over Japan. Intelligence analysts promptly dropped their sanguine assessment of the threat. "The probability that a missile armed with [weapons of mass destruction] would be used against U.S. forces or interests," a CIA-sponsored study asserted last year, "is higher today than during most of the cold war and will continue to grow."

The logic of the threat is simple. If we take North Korean, Chinese, and Iranian officials at their word, American "hegemony"—and, in particular, America's overwhelming military superiority—represents the single greatest challenge to their security. But, as the [1991] Gulf war showed, the United States can't be deterred with conventional forces alone. Ballistic missiles, by contrast, have proved they can do the job. There is, to begin with, the example of the Soviet Union, whose ICBM arsenal for decades kept the United States from confronting Soviet forces directly. More recently, North Korea's nuclear and missile programs have enabled that shambles of a country to blackmail the West into showering it with blandishments and concessions. Likewise, senior Pentagon officials say that China's repeated offers to incinerate Los Angeles linger in their calculations over how to respond to a conflict in the Taiwan Strait.

Nor have these examples been lost on states rushing to acquire long-range missiles: Their mere possession will put these countries "off limits" for U.S. intervention. "If [Americans] know that you have a deterrent force capable of hitting the United States, they would not be able to hit you," Muammar Qaddafi declared after the United States bombed Libya [in

1998, following a terrorist bomb attack at a disco in Germany that killed two U.S. soldiers]. "Consequently, we should build this [missile] force so that they and others will no longer think about an attack." Indeed, facing a dozen little Soviet Unions with even a theoretical capability of hitting America or her allies, the United States is vastly less likely to pursue a forward-leaning foreign policy. "[A]cquiring long-range ballistic missiles armed with a weapon of mass destruction probably will enable weaker countries to do three things that they otherwise might not be able to do: deter, constrain, and harm the United States," Robert Walpole, the national intelligence officer for strategic and nuclear programs, told the Senate Foreign Relations Committee in 1999. If, for instance, [Iraqi leader] Saddam Hussein possessed even one ICBM, U.S. forces wouldn't be bombing Iraq routinely or stationing troops nearby. "The idea is to keep us out of [an opponent's] neighborhood and prevent us from coming to the assistance of our allies—it'll work, too," says a senior administration official.

Missile defense is about preserving America's ability to wield power abroad.

Hence, when a missile defense opponent like Robert Reich writes in *The American Prospect* that the Bush team's plan exemplifies an "America-first policy" and "the new insularity," he has things exactly backward. The real argument for missile defense is that we need it to prevent adversaries from deterring us from the kind of interventions that liberals like Reich, even more than conservatives, spent the 1990s championing. Oddly enough, foreign critics, who carp that missile defense will cement U.S. hegemony and make Americans "masters of the world," grasp its rationale better than critics here at home. Missile defense, China's ambassador to the U.N. Conference on Disarmament complained recently, would grant the United States "absolute freedom in using or threatening to use force in international relations." He's right.

New dangers

The fact that ballistic missiles, as a 1999 National Intelligence Estimate points out, "are not envisioned at the outset as operational weapons of war, but primarily as strategic weapons of deterrence and coercive diplomacy" points to another flaw in the anti-ABM argument. Missile defense opponents argue that long-range missiles (and defenses against them) have become passe since rogue regimes surely intend, as Bill Clinton's National Security Adviser Sandy Berger put it in a recent *Washington Post* op-ed, to deliver "weapons of mass destruction by means far less sophisticated than an ICBM: a ship, plane or suitcase." Maybe. But a suitcase makes for much less menacing satellite imagery than an ICBM—which is to say, it has virtually no worth as a deterrent, much less any domestic political utility. Besides, if ballistic missiles are yesterday's news, then why are the North Koreans and the Iranians building them in the first place? "How about we get rid of our aircraft carriers and B-52s while we're at it?" scoffs a senior Bush adviser when confronted with the man-in-the-van argument. "You defend against what you can, and to argue that these missile programs aren't threats, or that because there are other threats we should ignore this one, is just silly."

Foreign critics, who carp that missile defense will . . . make Americans "masters of the world," grasp its rationale better than critics here at home.

Opponents of missile defense also rely heavily on McNamara-era logic. As Berger puts it, "[T]he basic logic of the ABM [Treaty] has not been repealed—that if either side has a defensive system the other believes can neutralize its offensive capabilities, mutual deterrence is undermined and the world is a less safe place." He's half right. If the United States fields missile defenses, *mutual* deterrence will indeed be undermined. In fact, it will be entirely one-sided in America's favor. Not only would a credible missile defense system diminish the ability of rogue states to deter the United States, but, because these states have so few missiles, even a limited defense would, if anything, diminish their confidence in their arsenals. Deterrence and missile defense may have been inherently incompatible when the United States faced an adversary armed with 9,000 warheads. But when the point is to deter a group of states that, between them, possess fewer than two dozen ICBMs, enshrining defenselessness in official policy makes no sense.

The difference between an adversary armed with a single warhead and one armed with 9,000 isn't the only distinction critics like Berger refuse to grasp. They also fail to consider how proliferation undermines cold war deterrence theory. If MAD, as Henry Kissinger has written, was "barely plausible when there was only one nuclear opponent," it's certainly less so today. That's because, in an era of proliferation, the numbers have become much less favorable for the United States. Instead of betting that one adversary will think like Berger, we are now pinning our survival on the hope that six or seven will.

Which brings us to the nature of those adversaries, a subject about which the anti-missile-defense lobby is remarkably sanguine. "[E]ven fanatical, paranoid regimes are deterred by the prospect of catastrophic consequences," Spurgeon Keeny, then executive director of the Arms Control Association, advised in a 1994 *New York Times* op-ed. Never mind that recent history is littered with paranoid regimes that forgot to be deterred by catastrophic consequences. Before we send Keeny to hammer out a SALT [Strategic Arms Limitation Talks] accord with Saddam, his ilk need to explain much more convincingly how missiles transform Third World dictators into rational choice theorists. Writing of "the 'psychological' element in deterrence, on which all else depends," Jonathan Schell, dean of nuclear abolitionists, notes that a leader's "state of mind—his self-interest, his sanity, his prudence, his self-control, his clear-sightedness—is the real foundation of his country's and everyone else's survival. In short, he must decide that the world he lives in is not one in which aggression pays off." Sanity, prudence, and self-control, needless to say, are not the first qualities that leap to mind when you think of leaders like Saddam and Qaddafi.

In any case, you don't have to be paranoid to miss the logic of MAD. Simple miscalculation will do.

Foreign relations concerns

Of course, there's more to U.S. foreign policy than relations with Iraq and other rogues. And, sure enough, the contention that missile defense will imperil America's relations with everyone else has become a favorite cliché of the anti-missile-defense chorus. The claim, though, is sheer invention. The Europeans have already rolled over. During his recent [2001] visit to the Continent, Defense Secretary Donald Rumsfeld brusquely told them that America was going to build an ABM system and there was nothing they could do about it. In the weeks since, officials from the European Union, Britain, Germany, and even France have lined up to declare that the United States has the right to deploy. In fact, serious European resistance has all but collapsed.

Russia, too, has nothing to worry about—and its officials know it. Unlike their predecessors in the 1980s, today's proposed missile defenses, which are being designed to intercept a much smaller number of warheads, pose no threat whatsoever to Moscow's huge arsenal. Indeed, [Russian president] Vladimir Putin ha[s] even proposed joining forces with the West to build missile defenses against rogue states. For all their bluster, German Foreign Minister Joschka Fischer said on a visit to Moscow, "[i]n the end, the Russians are going to accept it." As for the ABM Treaty, the other signatory—the Soviet Union—no longer exists. And, even when it did, it never paid the treaty much heed. In fact, the Soviets ringed the country with anti-missile systems, which still shield Russia today. That doesn't mean the accord should necessarily be abandoned. But neither should U.S. policymakers grant Russia veto power over America's ability to defend itself against unrelated threats.

If . . . the strategic rationale for missile defense really is an internationalist one, then a sea-based system has all the advantages.

As for the suggestion that missile defense will provoke China into what a recent petition by American sinologists described as "negative steps that would undermine American security," it's too late. Those "negative steps"—including unchecked missile proliferation and an arms buildup—have been under way for over a decade. Indeed, just two weeks ago U.S. bombers had to avoid hitting Chinese personnel working to upgrade Iraq's air defense systems. As the NSC's Joseph wrote, "China is modernizing its missile and nuclear arsenal whether or not the United States deploys missile defenses." And members of the Bush team contend privately that China's exports of missile technology and the expansion of its own missile program helped create the imperative for missile defense in the first place. Even so, ICBMs aren't free, and China has only about 20. Multiplying that arsenal several times over would require huge trade-offs. "If they build up aggressively," argues a senior Bush administration official, "there goes their trade relations with us, their multilateral diplomacy, maybe even their economy."

What kind of system?

The decision, in any case, has already been made—first by the Clinton administration and now by the Bush team: The United States is going to build a missile defense system. The question that matters is no longer if but how. Alas, here too the discussion has been a national embarrassment. Even though the systems under review today are limited and based on land and at sea, opponents of missile defense are still railing about what *Washington Post* hysteric Mary McGrory has revealed as "Bush's grandiose scheme for a real, all-out Star Wars scenario." Meanwhile, the right, too, has shown little interest in debating the competing merits of land- and sea-based missile defenses. Its approach is, instead, faith-based: Build it and it will come.

Within the U.S. government, however, a serious debate is under way. The ground-based option, slated for construction on a desolate Alaskan island, has the momentum. Unfortunately, it has few of the merits. In fact, all it has going for it is the ABM Treaty, which prohibits sea-based national missile defenses. To comply with the accord, the Clinton administration originally planned to erect a missile defense platform in North Dakota. Placing the system there had only one drawback: It offered protection to the continental United States but left parts of Alaska and Hawaii to fend for themselves. Ted Stevens, Alaska's senior senator and, more importantly, the head of the Senate Appropriations Committee, wasn't going to leave his constituents vulnerable to ballistic missile attack. So he went ballistic himself. By the time he finished, the Clinton White House had decided to relocate the site to Alaska.

The program, though, is a mess. First, it won't work. The problem isn't so much its well-publicized test failures (what failed in the most recent test was 50-year-old rocket technology that even North Korea has mastered) but a combination of flaws inherent in its design. The Alaska program can only intercept missiles well into their flight trajectories—that is, as they close in on the United States at a speed of about 15,000 miles per hour. Hence, the system would have only one shot at an incoming missile. Worse, that shot would likely have to maneuver through a cloud of decoys and countermeasures that ICBMs can deploy en route. Finally, the Alaska site will function reliably only against missile threats from East Asia. But a missile launched from, say, Iran or Iraq would be coming from the other direction. Given adequate time and resources, American technicians may solve these problems. Yet there's a conceptual defect they can never remedy—namely, that a U.S.-based missile defense amounts to just that. It abandons America's allies to their fates, offering Americans marginal protection but leaving countries like Israel and Japan defenseless.

A sea-based system

In truth, few experts champion the Alaska program. One of the reasons derives from the realization—arrived at rather late in the debate—that water covers 70 percent of the Earth's surface. This simple fact has enormous strategic and technical implications. First, it offers a way around the problem of decoys and multiple warheads. Missile interceptors stationed on U.S. ships, which would patrol the coasts of rogue states, could shoot down missiles in their initial boost phase—that is, before they could deploy countermeasures and while their engines were still emitting an easily detectable plume of flame. Trying to destroy a missile as it lifts off, as opposed to when it's about to land on you, makes sense on several counts. Aside from solving the countermeasure dilemma, it's a lot easier to hit. As anyone who has seen a televised space launch knows, rockets travel relatively slowly during their initial ascent—much more slowly than when they streak back to Earth.

Missile defense should respond to strategic imperatives, not political ones.

Equally important, a sea-based defense would offer the United States more than one opportunity to bring down an incoming missile. And, as physicist Richard Garwin points out, "It is much easier to put a lid on North Korea, a country the size of Mississippi, than it is to put an umbrella over the whole of the United States." If a missile did get through the first line of defense—or if it was launched from, say, China's vast interior, which no boost-phase interceptor could reach in time—American forces could conceivably have as many shots at it as there were ships stationed along the weapon's trajectory to the United States.

And, unlike a U.S.-based umbrella, a sea-based system wouldn't exclude America's friends. The mere fact that missile defense ships could be deployed to war zones as part of larger naval armadas gives them an immediately recognizable offensive dimension. Like aircraft carriers, such ships could project power in ways no concrete slab in Alaska could. If, as the Bush team insists, the strategic rationale for missile defense really is an internationalist one, then a sea-based system has all the advantages. The Alaska program, by contrast, follows the minimalist logic of Fortress America—that, as Alaska's Stevens wrote last month, "We can and we must defend our homeland!"

There's also a cost-benefit calculation. According to Congressional Budget Office estimates, the total cost of a ground-based program could run $60 billion. By contrast, the Pentagon has put the price tag of a sea-based defense at between $16 billion and $19 billion, while others have put it at half that. To be sure, when the cost of satellite-based sensors is factored into the equation, those projections may end up being too optimistic. Still, a sea-based system would build on an existing program: the Navy's Aegis air defense system, which has already been funded to the tune of $50 billion. And both the Clinton administration and the Pentagon report that one could be built with existing technologies.

What—you thought we were decades, if not centuries, from possessing the know-how to make missile defense work? In truth, NMD technology has matured well beyond what its detractors admit. The United States already fields a theater missile defense system, and current proposals aren't nearly as ambitious as the space-based plans of the Reagan era. In fact, the

Clinton administration purposely slowed down the Aegis's interceptor rocket to ensure it would not be usable against ICBMs and thereby violate the ABM Treaty. And, as boost-phase proponents (and prominent 1980s Star Wars critics) like Garwin and MIT technology Professor Theodore Postol point out, because this type of missile defense system targets rockets while they are still moving slowly and before they can deploy countermeasures, it's far less daunting technically than the Alaska program.

Analysts estimate that, with a new, more powerful missile interceptor and other upgrades to Aegis cruisers, the United States could begin deploying a sea-based defense in about seven or eight years. That's too long for some. President Bush has pledged to construct a system "at the earliest possible date." And a popular consensus has emerged that an Alaska-based defense could be completed more quickly. Several Republican senators, as well as representatives from the Pentagon's Ballistic Missile Defense Organization who have been working for years on the ground-based plan, favor continuing the Clinton program.

Until recently, the Bush team argued otherwise. Rumsfeld's chief of staff, Stephen Cambone, and Deputy National Security Adviser Stephen Hadley publicly derided the Clinton program as ineffective. And Joseph, Bush's NSC missile defense point man, wrote that the Alaska program "has become so contrived that it will have only a minimal capability against near-term threats." Even Bush called it "flawed."

No free lunch

But that was then. According to members of the Bush team and senior Pentagon officials, the White House is now considering proceeding with at least an Alaska-based radar system and possibly more. "Sea-based is unquestionably the better option, and we're going to pursue it," says an administration official. "But there's been a lot of work done on [Alaska], none done on Aegis, and we may end up doing both. Also, it gives us near-term insurance against North Korea." Two missile defense systems, of course, offer more protection than one. But in practice building anything more than a radar facility in Alaska would exact a high opportunity cost. To begin with, it would divert resources from the more promising system. In return, it would achieve minimal, if any, gains for U.S. security—and none for America's allies.

Equally important, missile defense funding comes from the military budget. But, as it stands, the military is already underfunded by about $30 billion annually, according to the Congressional Budget Office. Exactly how the Bush team would fund not one but two missile defense systems while rebuilding America's conventional forces remains anyone's guess. "There's no free lunch," says Michael Vickers, a military budget expert at the Center for Strategic and Budgetary Assessments. "NMD is an unfunded mandate, and you just can't do missile defense and much else at current costs." One solution, of course, would be to boost defense expenditures. But Bush insists there will be no new money for defense this year, and he has proposed a $1.6 trillion tax cut, which ensures there won't be much in the future either.

So where will the money come from? The Bush team says it plans to cut several major weapons programs and streamline the Armed Forces. Alas, the cuts Bush advisers privately suggest won't free up nearly enough money to fund their missile defense proposals, much less pay for their modernization plans. And there's only so much fat they can cut before they start bleeding the military's capacity to fulfill America's global commitments. Absent a substantial hike in defense spending, something has to give. Otherwise, America will have purchased an opportunity to wield its power undeterred at the expense of its actual capacity to do so. Which, needless to say, undermines the entire strategic rationale for missile defense.

Does Bush understand any of this? Probably not. But his advisers certainly do. They've spent years arguing that missile defense should respond to strategic imperatives, not political ones. Yet if they deploy a system whose purpose is to beat the clock rather than the missile threat, they will have done exactly what they've argued against for so long. The result will be a defense that encourages retrenchment while offering no security to our allies and very little to us. And, leftist critiques notwithstanding, that could be worse than none at all.

14

Pursuing a Missile Defense System Might Help Terrorists

Robert Wright

Robert Wright, a visiting scholar at the University of Pennsylvania, is the author of The Moral Animal *and* Nonzero: The Logic of Human Destiny.

Pursuing a missile defense system would likely provoke other countries into expanding their nuclear arsenals, which in turn may make it easier for terrorist groups to obtain nuclear weapons-grade materials. Thus, missile defense may actually increase the chances of another terrorist attack even more lethal than the one on September 11, 2001.

Some critics of President George W. Bush's missile-defense plan have already claimed a measure of vindication in Tuesday's [September 11, 2001] attack on the United States. Their logic is plausible as far as it goes: They've long been saying that massive destruction, when it comes, won't come via ballistic missile, and sure enough it didn't. But by ending their argument there, they're selling themselves short. Missile defense won't just fail to stop the next big terrorist attack. It could hasten the next attack and make it literally 100 times as lethal as Tuesday's.

Two questions

The question of whether terrorists might detonate a nuclear bomb in an American city has always boiled down to two questions: 1) Are there groups with significant resources and organizational skills that want to kill vast numbers of Americans? 2) How easy would it be for such a group to get ahold of the requisite materials? Tuesday gave us the grim answer to the first question. The answer to the second question is more elusive, but this much is clear: The more nuclear materials there are floating around beyond American control, the worse things look. And missile defense would probably raise that amount.

Missile defense won't just fail to stop the next big terrorist attack. It could hasten the next attack.

Both Russia and China have made noises about escalating their nuclear programs in response to missile defense. In the case of Russia, the threat rings hollow for fiscal reasons, but China has the resources to deliver. In fact, it is modernizing its arsenal in any event and can well afford to accelerate and expand the program in response to missile defense. And most experts agree that, within the framework of nuclear deterrence, doing so would be rational. In the more distant future, rapid growth in the Chinese arsenal could spur growth in India's arsenal, which could spur growth in Pakistan's arsenal, which could spur more growth in India's arsenal, and so on—with each iteration upping the chances of a little plutonium or uranium straying into the hands of terrorists.

The Bush administration seems to think that provoking the production of weapons-grade materials beyond America's borders is a fair price to pay for missile defense. In August 2001, the *New York Times* reported that the administration, to get Chinese acquiescence in the missile-defense program, had decided not to oppose China's nuclear modernization plans. This story, based on background reporting, was followed by hasty on-the-record qualifications and quasi-denials—whose upshot, as I read them, was to confirm the essential accuracy of the *Times* report.

The near indifference of many missile-defense boosters to nuclear proliferation has long been painfully clear. A few months ago, Charles Krauthammer of the *Washington Post*, in a column endorsing missile defense and the larger Bush vision that it's part of, celebrated the president's aversion to arms-control treaties. "Nor does the Bush administration fear an 'arms race.' If the Russians react to our doctrine by wasting billions building nukes that will only make the rubble bounce, let them." In the wake of what happened Tuesday, is Krauthammer still sanguine about the prospect of weapons-grade plutonium being produced not terribly far from Afghanistan, Iraq, and Iran, in factories run by underpaid bureaucrats? In any event, it's too late for him to retract his admission that Bush's policy could well lead to that. . . .

Unbalanced allocation of resources

Whenever you point out that a nuke is much more likely to enter the United States on a barge floating up the Hudson or a van crossing the Rio Grande than on a missile, missile-defense backers reply that we're already working on that problem. Yes, we are. But the point is that our allocation of resources is grossly unbalanced. The president plans to spend a massive amount of money to thwart incoming nuclear missiles, and less money to thwart nukes snuck across the border, a threat that everybody—*everybody*—considers more likely than a missile attack.

How exactly do we correct the imbalance? In theory, you should shift resources from missile defense to the fight against massively lethal terrorism until you're getting the same amount of protection per dollar from the two kinds of expenditure. Of course, we can't precisely quantify the "amount of protection" we get from each dollar spent to inhibit the spread of nuclear materials and other ingredients of mass murder. But the more likely these ingredients are to be acquired and used by terrorists, the more protection we're buying by slowing their spread. It's hard to reflect on what happened Tuesday without acknowledging that a major shift of resources is in order.

Still, some will manage. They'll argue—as if financial resources were infinite—that we can afford to keep the unprecedentedly expensive missile-defense program on track and *still* do more to cut the chances of a massive terrorist attack. But once you realize that ending or slowing the missile-defense program would itself cut the chances of such an attack—by cutting the world's incentive to produce nuclear materials—this argument starts to show signs of serious strain.

15

Ballistic Missile Disarmament Can Make Missile Defense Obsolete

Jurgen Scheffran

Jurgen Scheffran is a physicist at the Technical University Darmstadt, Germany, and is a cofounder of the International Network of Engineers and Scientists Against Proliferation (INESAP).

Ballistic missiles are costly and destabilizing weapons. Instead of attempting to create a technological defense system against them, the United States should work with other nations to eliminate such missiles through international disarmament agreements. America should begin with a moratorium on all testing and deployment of ballistic missiles and missile defense systems.

More than 15 years ago, in spring 1985, I had the opportunity to attend a meeting on the US Strategic Defense Initiative (SDI) near Munich in South Germany. Among the participants were several German officials, and as a representative of the US administration Paul Nitze was connected to part of the meeting via satellite link. While most questions to him were related to the implications of SDI for US-European relations, I made the following point. In his March 1983 speech (called Star Wars speech) US President Ronald Reagan had called upon scientists to develop technical measures to make nuclear weapons "impotent and obsolete". I asked Nitze: Wouldn't it be better to make SDI obsolete by nuclear and missile disarmament? I don't remember exactly Nitze's words, but in general he agreed even though he found the idea politically unrealistic.

In the following year, the Reykjavik summit between Reagan and the new Soviet leader Mikhail Gorbachev, ended with a big surprise. As both sides admitted, they had talked about eliminating all ballistic missiles. An agreement, however, failed because of disagreement on SDI. Gorbachev argued that SDI would destabilize the disarmament process. Ironically this event seemed to prove Nitze's assertion that comprehensive disarmament was unrealistic at that time, but the main reason was SDI.

The years after Reykjavik saw dramatic changes. As a consequence of Gorbachev's Perestroika, both superpowers agreed on initial disarmament steps in the Intermediate Nuclear Forces (INF) and Strategic Arms Reduction Treaties (START I and II). The Berlin Wall fell, the Cold War ended and the Soviet Union collapsed. The tremendous political changes seemed to make the SDI program obsolete, which also was suffering from technical flaws. Many observers supposed the era of confrontation had passed and hoped for a peace dividend. Even Paul Nitze seriously discussed the idea of Zero Ballistic Missiles (ZBM) during a meeting of the Federation of the American Scientists (FAS) in 1992.

We know that since then the world has changed again. The nuclear emperor has struck back and effectively prevented comprehensive nuclear disarmament. Thus it was not surprising that in the aftermath of the Gulf War, the missile defense idea was revived and will once again dominate the international security debate. The fact remains that because of technical difficulties, the nuclear emperor is still without defensive clothing. The argument that nuclear disarmament would make missile defenses obsolete is still valid. In April 1995, at the Review and Extension Conference of the Non-Proliferation Treaty (NPT) in New York, most governments and non-governmental organizations agreed that nuclear weapons should be eliminated, but there was little attention to the disarmament of delivery systems, ballistic missiles in particular.

Reasons for international missile control

There are good reasons for the strengthened international control and disarmament of ballistic missiles. These weapons allow aggressors to strike distant targets quickly, with little warning, and with a high probability of penetration. They played a destabilizing role and wasted enormous resources during the Cold War. Ballistic missiles are prestigious symbols of power, and the strongest reason to have them is that other countries do. The use of ballistic missiles in previous conflicts, from World War II to the Gulf War proved their political significance rather than their military utility. Even in the logic of deterrence and warfighting, ballistic missiles are obscure weapons. Their possession invites rather than deters attacks, and because of the extremely short warning times, they are highly destabilizing in a crisis. Since ballistic missiles are not a very efficient means of attacking military targets, they are largely used as weapons of terror against populated centers, based on the premise that attacks "out of the blue" frighten the people.

The argument that nuclear disarmament would make missile defenses obsolete is still valid.

With increasing range, ballistic missiles become excessively expensive. Because they fly through space, long-range missiles open the Pandora's box of space warfare. Paired with nuclear weapons the Inter-continental Ballistic Missile (ICBM) is the most terrible of all weapons, which is why missile proliferation has throughout history raised grave concerns, both regionally and globally. That even the US government (the largest missile power in the world) feels threatened about ballistic missiles of small countries proves too well that ballistic missiles do not provide security for anyone. On the contrary, they are the major cause for missile defense programs, which themselves are costly and destabilizing. Ballistic missiles are an especially threatening class of weapons and a world without them would be a better place for everyone.

Ballistic missiles are an especially threatening class of weapons and a world without them would be a better place for everyone.

There are several reasons why the control of these weapons should be improved. A focus on ballistic missiles should not exclude the control of other delivery vehicles, in particular cruise missiles and aircraft. It is better to have separate negotiating fora on the major means of delivery for biological, chemical and nuclear weapons. Accepting the argument of critics of ballistic missile disarmament that we need an all-at-once control of delivery systems would guarantee no progress at all. It would imply that the world is too complex to be changed, because everything is connected to everything.

There is still time to prevent a destabilizing and costly arms race between offensive and defensive missiles, assuming that the development of ICBMs is a complex and time-consuming task and the National Missile Defense (NMD) program takes longer than planned. In the past, ballistic missiles have been largely ignored in international arms control and disarmament negotiations, although the preamble of the NPT calls for "the elimination from national arsenals of nuclear weapons and the means of their delivery". In his speech to the House of Commons in London on July 3, 2000, Jayantha Dhanapala, the Under-Secretary-General for Disarmament Affairs of the United Nations, raised the question, "Why is public debate mired today in a duel between deterrence and defense, with scant attention to missile disarmament?"

The path towards missile disarmament

Previous efforts have focused on export control by the major suppliers of missile technology and bilateral arms control and disarmament of the former superpowers. The Missile Technology Control Regime (MTCR), now having 28 member countries, could delay some missile programs, but its effectiveness is limited by fundamental shortcomings. It is a voluntary, non-binding agreement with restricted membership and no enforcement, and it does not address already existing ballistic missile arsenals, ignoring the asymmetry between "haves" and "have nots".

To improve the present control regime, a few countries have made preliminary proposals within the limits of the MTCR. Some governments are now considering options for a stronger missile non-proliferation regime as an alternative to missile defense. Russia proposed a Global Control System for the Non-Proliferation of Missiles and Missile Technology

(GCS) that would increase transparency and reduce the risk of misunderstanding by notification of missile or space launches. To discourage proliferation, the GCS would offer security incentives and assistance in the peaceful uses of space. If the current missile owners would be allowed to keep their missile arsenals, then the effectiveness of a pure non-proliferation regime would be limited. The only way to deal with asymmetries between countries would be the creation of an international norm against all ballistic missiles.

In March 2000, ballistic missile experts from several countries met in Ottawa, Canada to explore options of a multilateral approach to more effective ballistic missile control, international monitoring, and early warning. The group saw a strong need to defend and expand the Anti-Ballistic Missile Treaty. To prevent instabilities and accidents, risk-reduction and confidence-building measures could be taken, such as de-alerting, improved ballistic missile early warning and launch notification. The concept of no-first use could be extended to ballistic missiles. For the long-term success of a missile control regime it would be important to "de-rogue" relations with countries such as North Korea and Iran and better understand their reasons for pursuing their missile programs. Recent political developments in these two countries have been rather positive in this respect.

A missile test ban would not be very difficult to verify.

According to the Ottawa Expert Group Report, the long-term goals include "demilitarization, the elimination of non-civilian ballistic missiles, and the elimination of nuclear weapons". A model for the elimination of ballistic missiles is the ZBM regime developed and discussed by the FAS in 1992. Such a regime would aim at the complete elimination of offensive ballistic missiles by a convention and combine unilateral declarations with regional and global multilateral agreements. The ZBM proposal suggested a step-by-step approach, including bilateral cuts between the USA and Russia, ballistic missile-free zones, an international Missile Conference, the creation of an International Agency for Ballistic Missile Disarmament, and finally agreement on the varying schedules to zero ballistic missile capability.

To ensure adequate verification, technical means of verification (sensors, intelligence) would focus on observable rocket characteristics (number, size, range, payload, deployment mode, launch preparations, flight trajectory). To prevent the transformation of space launch technology for ballistic missile purposes, cooperative verification and inspections, confidence building and data exchange is required. A safeguards system for space launchers could place some of the "most critical" items under supervision by an international organization. International cooperation in civilian space programs, removing incentives for national rocket programs, would be important in containing the use of space technology for missile development. A Canberra-style commission on "Cooperative Security in Space" could develop proposals to reduce military use of space and prevent the weaponization of space. [The Canberra Commission on the Elimination of Nuclear Weapons, an independent commission of global diplomats, was established in 1995 by the Australian government to identify steps to create a nuclear-free world.]

Act now

While global missile disarmament would be the longer-term perspective, the need for action is now. The best way to prevent an arms race and buy more time for political initiatives would be a moratorium on the further development, testing and deployment of ballistic missiles. A key element of such a missile freeze would be a ballistic missile flight test ban which would preclude the testing of new missiles, including missile defense systems, and reduce the chance of accidental or intentional war. A missile test ban would not be very difficult to verify since missile launches are visible from early warning satellites and ground- or air-based radars. Regional security initiatives could include the whole range of delivery systems to overcome asymmetries.

International organizations could facilitate such a process. Potential fora to discuss and negotiate multilateral missile control would be a conference of the MTCR member states and the Geneva Conference on Disarmament. Alternatively, an international conference of crucial missile countries could be considered.

Citizens and non-governmental organizations could play an important role in promoting and implementing missile control, in the context of the nuclear disarmament process. To increase public awareness, a greater public discourse on the missile problem and its resolution is needed. By building a network of information exchange and debate, experts, civil society and officials would be jointly engaged in discussing the broad aspects of the missile problem and its resolution. Activities could include meetings and conferences, involving scientists and technicians, as well as protesting and citizen inspections. Such endeavors are urgently needed to make missile defenses obsolete by disarmament before a new arms race can make disarmament impotent and obsolete.

Organizations to Contact

The editors have compiled the following list of organizations concerned with the issues debated in this book. The descriptions are derived from materials provided by the organizations. All have publications or information available for interested readers. The list was compiled on the date of publication of the present volume; the information provided here may change. Be aware that many organizations take several weeks or longer to respond to inquiries, so allow as much time as possible.

Arms Control Association (ACA)
1726 M St. NW, Suite 201, Washington, DC 20036
(202) 463-8270 • fax: (202) 463-8273
e-mail: aca@armscontrol.org • website: www.armscontrol.org

The Arms Control Association is a nonprofit organization dedicated to promoting public understanding of and support for effective arms control policies. ACA seeks to increase public appreciation of the need to limit arms, reduce international tensions, and promote world peace. Articles on missile defense appear regularly in its monthly magazine *Arms Control Today*.

Brookings Institution
1775 Massachusetts Ave. NW, Washington, DC 20036
(202) 797-6000 • fax: (202) 797-6004
e-mail: brookinfo@brook.edu • website: www.brookings.org

Founded in 1927, the institution conducts research and analyzes global events and their impact on the United States and U.S. foreign policy. It publishes the quarterly *Brookings Review* as well as numerous books and research papers on foreign policy, including the policy brief "Beyond Missile Defense: Countering Terrorism and Weapons of Mass Destruction" and the book *Defending America: The Case for Limited National Missile Defense*.

Cato Institute
1000 Massachusetts Ave. NW, Washington, DC 20001-5403
(202) 842-0200 • fax: (202) 842-3490
website: www.cato.org

The Cato Institute is a nonpartisan public policy research foundation that promotes the principles of limited government, individual liberty, and peace. The institute publishes policy analysis reports and op-eds, including "Go Slow on Missile Defense" and "What's the Right Missile Defense System for America?"

Center for Defense Information (CDI)
1779 Massachusetts Ave. NW, Suite 615, Washington, DC 20036
(202) 332-0600 • fax: (202) 462-4559
e-mail: info@cdi.org • website: www.cdi.org

CDI is comprised of civilians and former military officers who oppose both excessive expenditures for weapons and policies that increase the danger of war. The center serves as an independent monitor of the military, analyzing spending, policies, weapon systems, and related military issues. It publishes the *Defense Monitor* ten times per year.

Center for Nonproliferation Studies
Monterey Institute for International Studies
425 Van Buren St., Monterey, CA 93940
(831) 647-4154 • fax: (831) 647-3519
website: http://cns.miis.edu

The center researches all aspects of nonproliferation and works to combat the spread of weapons of mass destruction. The center produces research databases and has multiple reports, papers, speeches, and congressional testimony available online. Its main publication is the *Nonproliferation Review*, which is published three times per year. Among its reports is "The Impact of National Missile Defense on Nonproliferation Regimes."

Center for Security Policy (CSP)
1920 L St. NW, Suite 210, Washington, DC 20036
(202) 835-9077 • fax: (202) 835-9066
e-mail: info@centerforsecuritypolicy.org • website: www.security-policy.org

The center works to stimulate debate about all aspects of security policy, notably those policies bearing on the foreign, defense, economic, financial, and technology interests of the United States. It believes that the United States needs a robust missile defense system, and has produced numerous analyses and op-eds supportive of missile defense, most of which are available on its website.

Henry L. Stimson Center
11 Dupont Circle NW, 9th Fl., Washington, DC 20036
(202) 223-5956 • fax: (202) 238-9604
e-mail: info@stimson.org • website: www.stimson.org

The Stimson Center is an independent, nonprofit public policy institute committed to finding and promoting innovative solutions to the security challenges confronting the United States and other nations. Among its research projects is an assessment of American missile defense plans and possible reactions by China. The center produces occasional papers, reports, handbooks, and books on foreign policy and security issues.

Heritage Foundation
214 Massachusetts Ave. NE, Washington, DC 20002-4999
(202) 546-4400 • fax: (202) 546-8328
website: www.heritage.org

The Heritage Foundation is a conservative think tank that formulates and promotes public policies based on the principles of free enterprise, limited government, individual freedom, traditional American values, and a strong national defense. It publishes many position papers on nuclear weapons and missile defense, such as "Missile Defense: The Case Gets Stronger" and "Continue the Sea-Based Terminal-Phase Missile Defense Program."

Missile Defense Agency (MDA)
External Affairs
7100 Defense Pentagon, Washington, DC 20301-7100
(703) 697-8472
e-mail: external.affairs@bmdo.osd.mil • website: www.acq.osd.mil

The agency, formerly the Ballistic Missile Defense Organization (BMDO), is the establishment within the U.S. Department of Defense charged with responding to existing and emergent ballistic missile threats to the United States and its allies. Its website includes news on missile defense research, fact sheets on how ballistic missiles work, and congressional testimony by military leaders on missile defense.

Project Ploughshares
Institute of Peace and Conflict Studies, Conrad Grebel College, Waterloo, Ontario, Canada N2L 3G6
(519) 888-6541 • fax: (519) 885-0806
e-mail: plough@ploughshares.ca • website: www.ploughshares.ca

Project Ploughshares, an agency of the Canadian Council of Churches, promotes disarmament and demilitarization, the peaceful resolution of political conflict, and the pursuit of security based on equity, justice, and a sustainable environment. Public understanding and support for these goals is encouraged through research, education, and development of constructive policy alternatives. In articles published in the *Ploughshares Monitor* and elsewhere, it has opposed American plans for missile defense as being harmful for both Canadian and global security.

Union of Concerned Scientists (UCS)
2 Brattle Square, Cambridge, MA 02238
(617) 547-5552 • fax: (617) 864-9405
e-mail: ucs@ucsusa.org • website: www.ucsusa.org

UCS is concerned about the impact of advanced technology on society. It opposes deployment of missile defense. Publications include the quarterly *Nucleus* newsletter and reports and briefs on missile defense, including "Countermeasures: A Technical Evaluation of the Operational Effectiveness of the Planned US National Missile Defense System."

Washington File
U.S. Information Agency
301 Fourth St. SW, Room 602, Washington, DC 20547
(202) 619-4355
e-mail: inquiry@usia.gov • website: www.usia.gov

This website is a comprehensive source of current releases and government information relating to foreign affairs. It is maintained by the U.S. Information Agency, an independent foreign affairs agency within the executive branch of the U.S. government.

Bibliography

Books

James H. Anderson	*America at Risk: The Citizen's Guide to Missile Defense.* Washington, DC: Heritage Foundation, 1999.
Richard Butler	*Fatal Choice: Nuclear Weapons and the Illusion of Missile Defense.* Boulder, CO: Westview, 2001.
Anthony H. Cordesman	*Strategic Threats and National Missile Defense: Defending the U.S. Homeland.* Westport, CT: Praeger, 2002.
Frances Fitzgerald	*Way Out There in the Blue: Reagan, Star Wars and the End of the Cold War.* New York: Simon and Schuster, 2000.
Bradley Graham	*Hit to Kill: The New Battle over Shielding America from Missile Attack.* New York: Public Affairs, 2001.
Roger Handberg	*Ballistic Missile Defense and the Future of American Security.* Westport, CT: Praeger, 2002.
James M. Lindsay and Michael E. O'Hanlon	*Defending America: The Case for Limited National Missile Defense.* Washington, DC: Brookings Institution Press, 2001.
K. Scott McMahon	*Pursuit of the Shield: The U.S. Quest for Limited Ballistic Missile Defense.* Lanham, MD: University Press of America, 1997.
Gordon R. Mitchell	*Strategic Deception: Rhetoric, Science, and Politics in Missile Defense Advocacy.* East Lansing: Michigan State University Press, 2000.
Ben-Zion Naveh and Azriel Lorber, eds.	*Theater Ballistic Missile Defense.* Reston, VA: American Institute of Aeronautics and Astronautics, 2001.
James J. Wirtz and Jeffrey A. Larsen, eds.	*Rockets' Red Glare: Missile Defenses and the Future of World Politics.* Boulder, CO: Westview, 2001.
Earnest J. Yanarella	*Missile Defense Controversy: Technology in Search of a Mission.* Lexington: University Press of Kentucky, 2002.

Periodicals

Wade Boese	"The Pentagon Defends NMD Plans Amid Growing Skepticism," *Arms Control Today*, July/August 2000.
Steve Bonta	"The Case for Missile Defense," *New American*, December 3, 2001.
Business Week	"It's Rocket Science—and That's Not Good," May 14, 2001.
Philip C. Clarke	"Preventing Armageddon," *American Legion*, July 2001.

Congressional Digest "NMD in the United States: Evolution over the Last 50 Years," August/September 2001.

Philip Coyle "Rhetoric or Reality? Missile Defense under Bush," *Arms Control Today*, May 2002.

Charles Davis "Missile Defense Needed in a Dangerous World," *National Catholic Reporter*, February 1, 2002.

John M. Deutch et al. "National Missile Defense: Is There Another Way?" *Foreign Policy*, Summer 2000.

Charles Ferguson "Sparking a Buildup: U.S. Missile Defense and China's Nuclear Arsenal," *Arms Control Today*, March 2000.

Tim Folger "Shield of Dreams: A Critical Look at the Science and Technology Required to Build an Antiballistic Missile System That Would Make the United States Invulnerable to a Missile Attack," *Discover*, November 2001.

Evan Gahr "We Surrender: Gloom and Doomers Have Never Wanted Missile Defense to Work," *American Enterprise*, April/May 2001.

James K. Galbraith "Missile Defense: A Deadly Danger," *Dissent*, Summer 2001.

Charles L. Glaser and Steve Fetter "National Missile Defense and the Future of U.S. Nuclear Weapons Policy," *International Security*, Summer 2001.

William D. Hartung "Star Wars Unbound," *Nation*, April 8, 2002.

Bhupendra Jasani "U.S. National Missile Defence and International Security: Blessing or Blight?" *Space Policy*, November 2001.

Jeane Kirkpatrick "Dump the ABM Treaty," *American Enterprise*, April/May, 2001.

David Krieger "Stopping the New Nuclear Arms Race," *Humanist*, March 2001.

Steven Lambakis "Space Weapons: Refuting the Critics," *Policy Review*, February 2001.

Li Bin, Zhou Baogen, and Liu Zhiwei "China Will Have to Respond," *Bulletin of the Atomic Scientists*, November/December 2001.

Paul Rogat Loeb "The Money Defense Shield: A Little Political Honesty Here, Please," *Sojourners*, November/December 2001.

Richard Lowry "Missile Defense: The Time Is Now—Stop Talking and Start Building," *National Review*, April 2, 2001.

James P. Lucier "ABM Now," *Insight on the News*, October 1–8, 2001.

Morton Mintz "Two Minutes to Launch," *American Prospect*, February 24, 2001.

John Newhouse "The Missile Defense Debate," *Foreign Affairs*, July/August 2001.

Peter Pae "Kill Vehicle Scores a Hit with Proponents of Missile Defense," *Los Angeles Times*, March 26, 2002.

Geov Parrish   "Missile Mania: Arms Reduction Doesn't Mask Race To-
                      wards Missile Defense," *In These Times*, December 24,
                      2001.

Pavel Podvig   "For Russia, Little Loss, Little Gain," *Bulletin of the Atomic
                      Scientists*, November/December 2001.

The Progressive   "The Folly of Missile Defense," June 2000.

Tom Sauer   "Wrong in Too Many Ways," *Bulletin of the Atomic Scien-
                      tists*, November/December 2001.

Steven I. Schwartz   "Don't Know Much About History," *Bulletin of the Atomic
                      Scientists*, July 2001.

Index